Palgrave Studies in Medieval and Early Modern Medicine

Series Editors

Jonathan Barry
Department of History
University of Exeter
Exeter, UK

Fabrizio Bigotti
Institute for the History of Medicine
Julius Maximilian University
Würzburg, Germany

CENTRE FOR THE STUDY OF
MEDICINE AND THE BODY
IN THE RENAISSANCE

The series focuses on the intellectual history of medicine as related to the philosophies, institutions, practices, and technologies that developed across Mediterranean and Western traditions, c. 500–1800. Partnered with the Centre for the Study of Medicine and the Body in the Renaissance (CSMBR), it seeks to explore the range of interactions between various conceptualisations of the body, including their import for the arts (e.g. literature, painting, music, dance, and architecture) and the way different medical traditions overlapped and borrowed from each other. The series particularly welcomes contributions from young authors. The editors will consider proposals for single monographs, as well as edited collections and translations/editions of texts, either at a standard length (70-120,000 words) or as Palgrave Pivots (up to 50,000 words).

Marsha Wübbels

Quantification and Measurement in British Health Advice, c. 1670–1830

Diet, Exercise, and Body Weight in Iatromechanics

palgrave
macmillan

Marsha Wübbels
University of Exeter
Exeter, UK

ISSN 2524-7387 ISSN 2524-7395 (electronic)
Palgrave Studies in Medieval and Early Modern Medicine
ISBN 978-3-032-32759-8 ISBN 978-3-032-32760-4 (eBook)
https://doi.org/10.1007/978-3-032-32760-4

This work was supported by the Wellcome Trust (212622/Z/18/Z).

This Palgrave Macmillan imprint is published by the registered company Springer Nature Switzerland AG.
The registered company address is: Gewerbestrasse 11, 6330 Cham, Switzerland

SANTORIO AWARD

FOR EXCELLENCE IN RESEARCH

Florescat Flos Sapientiae

Acknowledgements My research for this project was funded by the Wellcome Trust, as a Doctoral Studentship in Humanities and Social Sciences [212622/Z/18/Z], and a Master's Award in Humanities and Social Sciences [208531/Z/17/Z], and I am immensely grateful for their support. I am grateful also to the staff at Berry Bros. & Rudd in St. James's Street in London, who kindly allowed me access to their ledgers, in which the body weights of thousands of their eighteenth-, nineteenth-, and twentieth-century customers were recorded. Parts of this book have been presented at conferences and webinars, and I have received helpful feedback from many people. I would like to thank the members of the Centre for the Study of Medicine and the Body in the Renaissance (CSMBR) for their support. I had the opportunity to present papers at the CSMBR's Summer School in 2021, and again in 2022, as part of a webinar titled *Measuring Health and the Body: Size, Weight and Pulse*, which was organised with the CSMBR's support. It is also the CSMBR who invited me to publish this book as part of the Palgrave Studies in Medieval and Early Modern Medicine (PSMEMM), an opportunity for which I am immensely grateful. I would like to thank in particular CSMBR director and Santorio-expert, Fabrizio Bigotti, whose feedback, advice, and support have been invaluable. Writing this book would not have been possible without the support of my PhD supervisors, Sarah Toulalan and Hester Schadee, and it is them in particular that I would like to thank for their input, advice, and support. Sarah especially has been my mentor throughout my academic journey—thank you so much, Sarah, for all your help. I am grateful also for the very helpful feedback from my thesis examiners, Alun Withey and Karen Harvey, who have both made suggestions to improve this book. Finally, I would like to thank my family for all their support over the course of my academic endeavours. I am grateful for my wife Zoey, and my siblings, Sanne, Arthur, and Stella, for their words of encouragement and confidence in my abilities. I would like to dedicate this book to my parents, Rika and Kasper, to thank them for all their help.

Contents

CHAPTER 1

Presentation of the Theme and Historical Contextualisation

Diet, exercise, and body weight were increasingly popular topics in the medical literature of the eighteenth century. The influx of exotic foodstuffs, rapid urbanisation, and increasing availability of luxury goods fuelled fears that people were becoming increasingly unhealthy. Rich diets, sedentary lifestyles, and cases of extreme corpulency were becoming increasingly widespread over the course of the eighteenth century, if physicians' warnings are anything to go by. Bath-based physician George Cheyne (1671–1743) admonished his patients and his readers for their overindulgence in 'gobbling down the richest and strongest food', and Timothy Bennet (1703–unknown) lamented in 1734 that exercise had become much neglected in these 'modern times'.[1] Rich foods and long lie-ins were the primary causes behind the influx of cases of corpulency, Thomas Short (1690?–1772) was sure, and he was equally certain that 'no Age did ever afford more Instances of Corpulency' than his own.[2]

[1] George Cheyne, *An Essay of health and long life* (London: Printed for George Strahan, 1724), 115 Eighteenth Century Collections Online; Timothy Bennet, *An essay on the gout; in which a method is propos'd to relieve the hereditary, and to cure the acquired … Such as have been reliev'd, or actually cur'd of the Gout by it* (London, printed for Richard Ford, 1734), 31–32. Eighteenth Century Collections Online.

[2] Thomas Short, *A discourse concerning the causes and effects of corpulency: together with the method for its prevention and cure* (London; printed for J. Roberts, 1727), 10. Eighteenth Century Collections Online.

M. Wübbels, *Quantification and Measurement in British Health Advice, c. 1670–1830*, Palgrave Studies in Medieval and Early Modern Medicine,
https://doi.org/10.1007/978-3-032-32760-4_1

A plethora of rules, aphorisms, and general advice existed to combat these issues, published in the many guides to health that were written for all those willing to heed physicians' warnings. While such warnings may sound familiar to modern readers, I will argue in this book that the advice found in eighteenth-century medical advice is singular and the result of the coming together of a set of theories unique to this period. As such, it defies easy comparison with ancient or indeed modern health advice and is best understood with reference to a multitude of factors and developments which affected physicians' advice about diet, exercise, and body weight.

According to many authors of eighteenth-century health advice, a man was either 'a fool or a physician'. This common saying reflects the contemporary belief that health depended on people's knowledge of medicine and the functioning of the body. It is therefore perhaps not surprising that literature about health was a very popular genre in the eighteenth century.[3] Health advice was not limited to conversations with the physician but could be found in a wide range of medical literature, from the most academic of treatises to relatively cheap and popular texts promising easy-to-follow rules. This book explores health advice expressed in published health guides, pamphlets and medical treatises, as well as letters written between 1670 and 1780, and manuscript source material in the form of physicians' case notes, and ledgers detailing weighing practices from the mid- to late eighteenth century. Some material which appeared in print between 1670 and 1780 may have been written before 1670 but has nevertheless been included in the analysis due to its continued relevance. Furthermore, for the purpose of making comparisons, a number of publications that fall beyond these timeframes have also been included.

Within the topics of diet, exercise and body weight, sources dating from the period 1670–1780 stood out as evidence of significant developments. These dates coincide with the rise and fall in popularity of iatromechanism (sometimes also called iatrophysics or iatromathematics) as a school of thought that applied physics to the field of medicine. Medical texts written within this period suggest that iatromechanical approaches to

[3] Ginnie Smith, "Prescribing the rules of health: Self-help and advice in the late eighteenth century" in *Patients and Practitioners: Lay perceptions of medicine in pre-industrial society*, edited by Roy Porter Cambridge: Cambridge University Press (1985), 252–253; Paul Slack, "Mirrors of health and treasures of poor men: the use of vernacular medical literature in Tudor England", edited by Charles Webster. (Cambridge: Cambridge University Press 1979).

medicine and health significantly affected contemporary health advice. By the late eighteenth century, when the popularity of mechanical theories started to dwindle, aspects of health advice that can be attributed to the influence of mechanical ideas similarly started to wane, especially in discussions about diet and body weight. In discussions about diet, the mechanists' focus on quantitative food measuring and the shape and digestibility of food particles became replaced with chemical food analysis. This shift has also been noted by Elizabeth Williams and Rebecca Earle, who situate the beginning of our modern approach to diet in the late eighteenth and early nineteenth centuries. Discussions about fatness and body weight also changed in the late eighteenth century. As Chap. 5 will show, we encounter the first indications that corpulency became explicitly linked to body weight in the 1780s, although it is still not entirely clear why that is the case. Regardless of the causes, however, advice about diet and body weight clearly changes between the late eighteenth and early nineteenth centuries, and the chronological boundaries of this book reflect that. Discussions about exercise follow a slightly different chronology than those about diet. Aspects particular to eighteenth-century exercise advice, such as the preference for motions which stimulate the minute fibres of the body, and the fears around perspiration, continue into the early nineteenth century, although these concerns do seem to abate somewhat compared to early eighteenth-century texts. The continuity of these aspects could be related to shared theories between iatromechanism and vitalism. The source selection for the chapters about exercise demonstrates the continuities between early and late eighteenth-century discussions about exercise, and to show that change in this field occurred later, I shall here discuss medical texts beyond 1780, up to 1830.

The exact chronology of shifts in health advice and practices within this period can be difficult to pin down, though some narratives are clearly discernible. The medical school of thought known as iatromechanism, especially influential in the late seventeenth and early eighteenth centuries, was the impetus for the renewed interest in the theories and experimental practices of Paduan physician Santorio Santori (1561–1636). Santorio is perhaps most famous for his experiments, in which he measured and weighed himself and others daily over a long period of time. The results of these experiments were published in his most famous work, *Ars de statica medicina* (Venice, 1614). *Ars de statica medicina* (hereafter *Medicina statica*) consisted of a series of aphorisms arranged according to the six non-naturals (air, exercise, sleep and waking, food and drink, excretions, and the passions of the mind), with the addition of a chapter on insensible perspiration. It was in the chapter of insensible perspiration ('perspiratio insensibilis') especially that Santorio discussed the role of body weight in health maintenance. Santorio's experiments, designed to measure and manage what we would now call metabolism, were re-tried, republished, and re-discussed over the course of the early eighteenth century. The effects of these developments on dietary and exercise advice were significant. Chapters 2 and 3 of this book trace the introduction of iatromechanical and Santorian concepts and practices to dietary advice and argue that health guides published between the 1710s and 1740s increasingly emphasised the importance of dietary quantities to health maintenance and in some cases encouraged the measuring of dietary intake by weighing food or the body. As Chaps. 4 and 5 demonstrate, exercise advice of this period similarly underwent considerable changes following the popularity of the iatromechanical school of thought and the re-introduction of Santorio's work. Exercise enjoyed greater attention from health authors than it had ever before due to iatromechanics' concern with the solid, rather than the fluid parts of the body, and exercise advice too shows clear influences from Santorio's theories, especially relating to sweat and insensible perspiration. The chronology of the practice of and theory relating to body weighing is less neat, though this too can be linked to the rise and fall of the popularity of the iatromechanical school of thought. Between the 1710s and 1740s especially, the topic of body weighing enjoyed great interest and sparked much debate among physicians. However, as Chap. 6 explains, the practice of weighing oneself did not spread beyond medical circles until after the 1760s, when interest in iatromechanism, and Santorio, had already started to wane. Technological developments can perhaps explain the delay between medical and popular practices, as evidence indicates that public weighing machines started to become available in Britain from the 1760s, and personal weighing machines not until the 1780s. Even so, the influence of Santorio is still palpable in sources which document late eighteenth-century weighing practices, as will be shown.

The eighteenth century was a period in which many medical schools of thought co-existed. Hippocratic and Galenic theories still found expression in this period but increasingly were challenged by not only iatromechanical, but also vitalist ideas. Although this book takes into consideration the effects of all these schools of thought on health advice, it focusses especially on the effects of iatromechanical theories on discussions of diet, exercise, and body weight. The purpose of this particular focus is twofold:

firstly, it is the contention of the author that iatromechanism had a particularly significant effect on health advice, especially relating to the topics of diet, exercise, and body weight, and that these effects have not been sufficiently explored in historical research so far. Secondly, this study aims to fill a gap in the historiography. The iatromechanical school of thought and its effect on the medical literature of the eighteenth century remains relatively unexplored compared to iatrochemistry or vitalism.[4]

Iatromechanism became popular in the medical circles of the late seventeenth century, following the works of scientists and physicians such as René Descartes (1596–1650), Giovanni Alfonso Borelli (1608–1678), Marcello Malpighi (1596–1650), and Archibald Pitcairne (1652–1713).[5] Iatromechanical theories were inspired by scientific discoveries such as William Harvey's (1578–1657) circulation of the blood and Antoni van Leeuwenhoek's (1632–1723) blood corpuscles.[6] Particularly important to this school of thought was the work of Santorio, whose theories and experimental methodologies were appropriated by eighteenth-century mechanically inclined physicians such as John Floyer (1649–1738), George Cheyne (1671–1743), and James Keill (1673–1719), but especially apothecary and medical writer John Quincy (unknown–1722).

Quincy claimed that Santorio's work was already 'a particular Favourite with the Faculty' when he published his first translation of Santorio's *Medicina Statica* in 1712. The aim was therefore not to re-introduce this work into medical writing but to bring it 'into a larger Acquaintance' by translating it into English and adding explanations, rendering the text 'easy and intelligible, almost to any Person', Quincy explained in his Preface.[7] It was not the first time *Medicina statica* had been translated into English: in 1676, Santorio's aphorisms were published as *Rules of Health,* but Quincy argued that version to contain too 'many Terms and Latin Phrases', limiting its potential readership.[8] Another version of *Medicina*

[4] As has also been argued by Hisao Ishizuka, *Fiber, Medicine, and Culture in the British Enlightenment* (New York, Palgrave Macmillan, 2016).

[5] A helpful overview of the shifts and developments of medical theories and schools in the early modern period, including the development of the iatromechanical school of thought, can be found in Ruben E. Verwaal, *Bodily Fluids, Chemistry and Medicine in the Eighteenth-Century Boerhaave School* (London: Palgrave Macmillan, 2020), 3–4.

[6] Ibid., 10.

[7] Quincy, Preface, *Medicina Statica* (1737 version), iii.

[8] Ibid., v; *Medicina Statica: Or, Rules of Health, in Eight Sections of Aphorisms.* Trans. J.D. (London, Printed for John Starkey, 1676).

statica, with added comments from physician Martin Lister (1638?–1712), was published in Britain in 1701, though this text was written wholly in Latin.[9] Quincy republished his translation another four times between 1720 and 1737, adding further comments, essays, and comparisons between the findings of Santorio and those of English experimenter James Keill.[10] It appears that efforts to popularise the *Medicina statica* paid off, as by 1711, the 'Learned World' was 'very well acquainted' with Santorio and his weighing chair, according to London politician Joseph Addison (1672–1719).[11]

Not only were Santorio's aphorisms republished in Britain, but they were also re-tried. Starting in 1708, physician James Keill undertook a weighing experiment in Northampton, to ascertain how perspiration levels were different in the English climate than they were in the Italian climate, where Santorio had undertaken his experiment.[12] He published the results of his experiments under the title *Medicina Statica Britannica* in 1717, and his work was re-published by Quincy in 1720, together with Quincy's translation of Santorio's *Medicina statica*. Similar experiments were undertaken in Ireland by physicians George Rye (between 1721 and 1722) and Bryan Robinson (between 1721 and 1745) and in Scotland by Edinburgh physician Francis Home in 1751.[13] The final experiment undertaken in Britain was concluded in 1769, by London physician William Stark. As Lucia Dacome has argued, these physicians engaged in a practice 'that they recognised as characteristically "Sanctorian"'.[14] The purpose of these experiments was not to call into question Santorio's overall arguments, but rather to make the aphorisms applicable to their own readership by adjusting for a colder, British climate, which Santorio himself had argued would affect the measurements. The re-trials therefore lend further credit still to Santorio's *Medicina statica*, as well as his

[9] Sanctorii Sanctorii, *De Statica Medicina Aphorismorum, Sectiones Septem: cum Commentario Martini Lister*. Trans. Martin Lister (London, Impensis Sam. Smith, 1701).

[10] Keill's *Medicina Statica Brittanica* was added to the second edition, published in 1720.

[11] Addison in *The Spectator*, Addison and Richard Steele, Vol. 1, No. 25 (March 29, 1711), 93. Available online at: <https://archive.org/details/in.ernet.dli.2015.39152/page/n101>. Accessed 19 January 2020.

[12] Keill, *Medicina Statica Britannica*, 'Dedication', unpaginated.

[13] Francis Home, *Medical facts and experiments* (Edinburgh, 1759); Dacome, "living with the chair", 487.

[14] Dacome, "Living with the chair", 467.

invention of the weighing chair as a legitimate method for measuring the health of the body.

The question of whether Santorio's weighing experiment could or should be undertaken by non-medical people as well was a topic of much debate in the wake of the re-trials. Some physicians, such as George Cheyne, recommended that readers tried the method of a 'Sanctorian Chair' to work out how much food they needed to manage their body weight, while others argued that such an experiment was not only tedious, but also unnecessary.[15] Joseph Addison was especially outspoken against the idea of ordinary people using Santorio's 'Mathematical chair', warning in 1711 in *The Spectator* that those who would live by 'Ounces and Scruples' instead of the 'natural solicitations of Hunger and Thirst' would temper their health, rather than keep it.[16] Evidence of non-medical people weighing themselves in a manner similar to Santorio remains scarce for this period, though Addison's article in *The Spectator* suggests that at least some social commentators seemed concerned that this practice was more widespread than is apparent. Though much of this discussion was confined to the early eighteenth century, the name 'Santorio' (often written in its Latin spelling, 'Sanctorius') and his invention can be found in much later texts as well. In 1786, Belgium-born inventor John Joseph Merlin (1735–1803) advertised his mechanical weighing chair as a 'Sanctorius's Balance' in newspapers and pamphlets, which indicates that even in the late eighteenth century, the practice of weighing oneself continued to be associated with Santorio and his experiment.[17]

Iatromechanism was a school of thought rather than one particular theory or set of ideas which shifted and developed over time and can therefore be difficult to summarise. One of the key aspects of this medical movement was the view of the body as a mechanical construction. Iatromechanists often described the body as a hydraulic machine consisting of a network of numerous narrowing tubes through which the blood,

[15] George Cheyne, *An essay on Regimen* (London, Dan. Browne, 1753), Iiii. Emphasis my own.

[16] Joseph Addison in *The Spectator*, a daily publication founded by Addison and Richard Steele, Vol. 1, No. 25 (March 29, 1711), 93. Available online at: <https://archive.org/details/in.ernet.dli.2015.39152/page/n101>. Accessed 19 January 2020.

[17] The exhibition was advertised in a pamphlet: John Joseph Merlin, *Morning and Evening Amusements at Merlin's Mechanical Museum, No. 11 Princess Street, Hannover Square* (London, 1787), https://archive.org/details/b30375642/page/6, 7; Merlin in *Morning post*, issue 4032 (London, Wed Jan 4th, 1786) advertisements and notices.

consisting of minute particles, flowed. Blood was propelled forth through the mechanical pumping motion of the heart, as well as the pressure provided by the tubes (often called vessels) themselves. Physicians of this school of thought ascribed much importance to the smooth circulation of fluids through the complex tubular network and any form of obstruction or weakening of this flow was considered an immediate and considerable threat to health. Another popular image for the body was that of a clock with all its minute cogs on which it relied for its ability to move. Whatever the metaphor, iatromechanical theorists sought to convey the idea that the body was an automaton, the functioning of which relied on the proper maintenance of every individual part, from the largest organs to the minutest vessels.

Many aspects of iatromechanism were a continuation of humoural ideas, with the terms used to explain these ideas signalling the difference between old and new understandings of health and the body. The mechanists' focus on the hydraulics of the body, for example, closely aligned to previous humoural concerns with the flow of the humours and the potential dangers of stagnation or blockages to health. According to classical medical theory, all bodies consisted of a mixture of humours in which generally one of the four humours (blood, phlegm, yellow bile, and black bile) dominated. Accordingly, there were many different 'constitutions', for various combinations of the dominating humours. Bodies which abounded in blood were generally said to be of a 'Sanguineous' constitution, bodies which abounded in Phlegm were 'Phlegmatick', while 'Cholerick' and 'Melancolick' bodies had a great proportion of yellow bile and black bile, respectively. Having a predominant humour was not necessarily indicative of ill health: all bodies had more or less of particular humours, and this did not generally cause problems. It was only when a humour abounded to the extent that it started to cause illness, that physicians recommended trying to re-balance the humoural state of the body through the management of the 'non-naturals'.

Galen argued that there were six major causes which could affect the humours: these were air, exercise and rest, sleep and waking, food and drink, secretions and retentions, and the passions of the mind. They were called the 'non-naturals' by Johannitius in his introduction to Galen's *Art of Medicine* to convey that these six factors were not internal to the body but rather derived from external causes such as the surroundings (especially air) and personal conduct (such as errors in diet, or excessive or insufficient exercise).[18] As different individuals had different constitutions and lived in different environments and had varying lifestyles, Galenic medicine was largely a matter of physicians carefully judging the best course of treatment for every individual patient. Such a course was called a 'regimen' and consisted of directions for the management of all the six non-naturals. Management of the non-naturals was generally a case of either preserving the humoural equilibrium if a person was healthy (by making no sudden alteration in one's dietary, exercise, or sleeping habits, and no change in environment) or correcting the state of the humours if a person was ill. Correction was generally pursued through administering treatments which were qualitatively opposite to the nature of the illness. For example, if a patient was suffering from the consequences of an excessively hot and dry constitution (choleric), a physician might prescribe him with food which was of a cold and wet nature to correct this imbalance. Humoural medicine was therefore quintessentially qualitative in its explanations of health and illness.

State of the Art

The iatromechanical school of thought envisioned the body and all its physiological processes to be mechanical and quantifiable, and the physicians who espoused these ideas argued that all bodily processes were explicable and predictable through measuring and mathematical calculations. Another aspect particular to iatromechanism is therefore the trust placed by physicians in measuring and mathematics as reliable tools for medical practice. In the early eighteenth century in particular, several physicians conducted experiments seeking to measure and quantify the body and the bodily process we now call metabolism, often using instruments which allowed for precision measuring. The term quantification is, as Simone Guide and Joaquim Braga have shown, difficult to define, as the meaning of the concept changed over time.[19] In this book, the term 'quantification'

[18] Sachiko Kusakawa, "Medicine in Western Europe in 1500," in *The Healing Arts, Health, Disease and Society in Europe, 1500–1800,* ed. Peter Elmer, (Manchester, Manchester University Press), 8.

[19] Simone Guidi and Joaquim Braga, "introduction" in Simone Guidi and Joaquim Braga (eds.), *The Quantification of life and health from the sixteenth to the Nineteenth century: Intersections of Medicine and Philosophy,* in Palgrave Studies in Medieval and Early Modern Medicine, Palgrave Studies in Medieval and Early Modern Medicine (London, Palgrave Macmillan Cham, 2023), https://doi.org/10.1007/978-3-031-15725-7.

will mean the act of expressing the body, bodily processes, and acts of health maintenance, in numbers, specifically for the purpose of establishing parameters for determining what is healthful and what is not. The quantification of medicine between the late seventeenth and late eighteenth centuries is noticeable especially in health advice relating to diet, exercise, and body weight. With precision scales, physicians could perceive and measure fluctuations in weight and use these figures to determine the proportions between the body's excretions (and therefore learn which excretions were the most significant), but also measure the effects of illnesses or non-naturals on the body's excretions. Food and drink were increasingly expressed in numbers as well. More than ever before, physicians warned readers of health guides to mind the quantity of food they ate, and a few physicians went so far as to quantify the exact measurement of food and drink they believed would suit most people. Daily exercise also increasingly became expressed in numbers: a few individuals weighed their bodies before and after exercise to find out how much sweat was lost through exercise, while others quantified exercise through recommending a specific length of time one should spend exercising. Measuring, especially when using precision instruments, is what made quantifying the body and health more possible than it had ever been before: minute changes to our body weight caused by eating a meal or excreting through stool, urine, or sweat are not usually noticeable but can be perceived by us when these are indicated by a weighing instrument precise enough to indicate minute changes. It is therefore because of measuring instruments such as scales that physicians could perceive, and express, in numbers, excretions which would otherwise remain imperceptible and difficult to compare and evaluate and quantify. Precision instruments allowed physicians to observe what could otherwise not be observed, establish parameters which were previously determined through the senses alone, and compare values within and outside the body and between bodies.

This aspect of iatromechanism has enjoyed particular interest from scholars of medical thought, resulting in a number of works of note. The earliest studies into the history of quantification in medicine are those of Owsei Temkin, who investigated the possible influence of Galen's quantitative observation of urine produced by the kidneys on William Harvey's (1578–1657) attempts to quantify the amount of blood propelled into the aorta in a specific time frame.[20] In 1977, Jerome J. Bylebyl continued the discussion of the link between Galen and Harvey's quantitative reasoning for Harvey's hypothesis that the amount of blood transmitted by the heart could not possibly be supplied through ingested food and that blood therefore had to circulate through the body.[21] Temkin and Bylebyl have both shown that quantitative reasoning existed in medicine long before the eighteenth century. It is in the seventeenth and eighteenth centuries however that we see a notable increase in quantitative reasoning and measuring experiments in the field of medicine and this is reflected in the recent scholarship. *The Quantification of Life and Health from the Sixteenth to the Nineteenth Century* (2023), edited by Simone Guidi and Joaquim Braga, discusses important moments in the quantitative trend in Western medicine between the sixteenth and nineteenth centuries through a set of case histories and illustrates well the complexities involved in understanding quantification as a concept.[22] The chapters in their volume discuss issues which were both of medical and philosophical import, such as the quantification of temperaments and complexions, the quantification of life processes and physiology, and the quantification of embryology, as well as the impact of quantification on notions such as health and illness, as well as virtue, vice, and personhood. Quantification in medicine is also the focus of *Santorio Santori and the Emergence of Quantified Medicine, 1614–1790: Corpuscularianism, Technology, and Experimentation* (2022), edited by Jonathan Barry and Fabrizio Bigotti, which contains extensive research on Santorio's quantitative research methods and measuring instruments, and chapters which explore the impact of Santorio's work on later physicians such as Daniel Sennert, Alfonso Borelli, Giorgio Baglivi, and Carl Linnaeus.[23] Sandra Sherman and Emma Spary investigated what they argue to be a rise in the use of quantifiable methods specifically for

[20] Owsei Temkin, 'A Galenic model for quantitative physiological reasoning? *Bulletin of the History of Medicine*, vol. 35, no. 5, 1961, pp. 470–75. *JSTOR*, http://www.jstor.org/stable/44446820; Owsei Temkin "Nutrition from classical antiquity to the baroque," in *Human Nutrition, Historic and Scientific*, ed. Iago Galdston (New York, International Universities Press, 1960).

[21] Jerome J. Bylebyl, "Nutrition, quantification and circulation," *Bulletin of the History of Medicine*, 51, 3 (1977), 369–385.

[22] Guidi and Braga, *The Quantification of life and health from the sixteenth to the Nineteenth century*.

[23] Jonathan Barry and Fabrizio Bigotti, *Santorio Santori and the Emergence of Quantified Medicine, 1614–1790: Corpuscularianism, Technology, and Experimentation* (London, Palgrave Macmillan, 2022).

determining sustainable dietary quantities, and historians such as Lucia Dacome, Michael Stolberg, and Alexander Pyrges have all looked at quantitative methodology in relation to medical discussions about body weight.[24] These scholars have done much to trace the quantification of medical research in the seventeenth and eighteenth centuries, and this volume builds on their findings.

This volume draws our attention to the effects of quantification on English health advice available to the public between the late seventeenth and early nineteenth centuries. It will explore the extent to which authors of health literature used quantification and measurement to explain the body, the bodily processes, and the extent to which they used measurement in their advice on how to manage the health of the body, especially relating to diet and exercise, and body weighing. As we shall see, quantification and measurement played a considerable role in these topics and understanding how these topics were influenced by the quantitative trend in medicine goes a long way towards explaining the significant differences between modern and early modern ideas on how to look after the body through diet, exercise, and body weighing. Where other projects have primarily focussed on the application of quantification methods in medical circles, this book takes this investigation one step further by assessing how physicians communicated these methods to the wider readership of their treatises. Historians of medicine have long debated the extent to which non-medical people engaged with medical theory and practice, and this investigation contributes directly to this ongoing debate through its assessment of physicians' considerations for their patients' unique bodies

[24] Sandra Sherman, *Imagining Poverty: Quantification and the Decline of Paternalism* (Columbus, Ohio State University Press, 2001); E. C. Spary, "Food," in *A Cultural History of Medicine: In the Enlightenment* ed. Lisa Wynne Smith (London, Bloomsbury, 2021); Lucia Dacome, "Balancing Act", *Studies in History and Philosophy of Biological and Biomedical Sciences*, 43, (2012); Lucia Dacome "Living with the chair: private excreta, collective health and medical authority in the eighteenth century," *History of Science*, 39 (2001); Lucia Dacome, 'Useless and Pernicious Matter: Corpulence in Eighteenth-Century England', in Cristopher E. Forth and Ana Carden-Coyne (eds.), *Cultures of the Abdomen: diet, digestion, and fat in the modern world* (Palgrave Macmillan, Hampshire, 2005); Michael Stolberg, "'Abhorreas pinguedinem': Fat and obesity in early modern medicine (c. 1500–1750)", *Studies in History and Philosophy of Biological and Biomedical Sciences*, 43 (2012), 370–378; Alexander Pyrges, "'No Age Did Ever Afford More Instances of Corpulency'. Obesity as a Collective Condition and the Early Modern Medicalization of Girth", in Jill Burke, Holly Fletcher, Christine Ott (eds.), *Fat Bodies in Early Modern Europe* (Forthcoming).

and lifestyles. To look for the patient through the eyes of the physician of course comes with its own challenges. The intention, however, is not to determine whether or not patients followed the advice of physicians, but rather to investigate how physicians ensured that patients could follow their advice. This book considers how medical theory was conveyed into practicable guidelines and gauges the reasons behind physicians' decisions to recommend particular methods of maintaining health.

This study limits itself to investigating medical texts written in the English language and published within England, although some Scottish sources are used as well, such as letters written by famed Edinburgh physician William Cullen, whose works were published all over Europe. One Irish and one Dutch source are used as well, on the basis that these were discussed by English physicians in English texts and therefore were part of the English debates. Also included are texts which were translated into English from Latin, especially where these enjoyed significant popularity in Britain. The purpose of this selection of sources is to give a wide impression of the discussions which took place within the British medical milieu. Health advice was generally addressed to an audience of a particular social rank and these sources must be understood to represent only people of that subset of society, and not everyone living in Britain in the long eighteenth century. The limitations as to who these sources represent relate not only to matters of social rank but also to demographics. As will be discussed extensively in the chapters themselves, physicians frequently discussed what they perceived to be threats to health common primarily, if not exclusively, to the urban population of Britain. These limitations bear little influence on the inquiries into the effects of iatromechanical theories on health advice. They do of course require consideration in relation to the second inquiry, namely physicians' considerations for their readership. As this project will demonstrate, the authors of health advice anticipated not only the social rank, but also the lifestyle and location of their readership, and based their advice on these assumptions. While the nature of these sources therefore does not afford us a full picture of the health practices of all, or even most British people, they do lend themselves well to an investigation into the considerations that eighteenth-century physicians made when constructing their advice.

This volume builds upon scholarship in the broader field of medicine, as well as on more specific work done in the histories of the body, dieting, exercise and sports history, and the history of health literature. Where this book breaks new ground is in combining the topics of diet, exercise and weighing. Research so far has tended to focus on either one particular non-natural—air, diet, exercise, excretions, sleep, and the passions of the mind—or all six at the same time. The treating of these three topics together is warranted because these three topics are considered intrinsically connected not only today, but also in the long eighteenth century. This continuing connection therefore begs the question of how it was understood in the eighteenth century and to what extent this differs from our understanding today. This book argues that the relationship between exercise, diet, and body weight was explained using very different theories than those today. Consequently, evidence of a relationship between these topics must not be misunderstood to mean that our modern ideas about diet, exercise, and body weight stem from eighteenth-century theories or that there is a direct linear development between them. It is the contention of this book that in order to understand eighteenth-century health concerns, we must disconnect our modern presumptions from what we read in these texts.

The longevity of Galen's doctrine of the six non-naturals was one of the first areas of disagreement among historians.[25] Since the 1960s, historians, starting with L. J. Rather, have studied the origin of the term non-naturals and their role in early modern medical literature.[26] Saul Jarcho called attention to the topic by documenting the discussions of these non-naturals in early modern texts, concluding that 'until the end of the eighteenth century every graduate physician was probably quite familiar with Galen', to the extent that he found it was customary to mention the non-naturals without needing further reference.[27] Peter Niebyl further argued that the non-naturals remained central to early modern medical literature,

[25] Ginnie Smith pointed out the importance of the regimen genre to historians of medicine, noting that they were an extremely popular genre at the time, and argued that they would have left few sections of the population untouched. Ginnie Smith, "Prescribing the rules of health: Self-help and advice in the late eighteenth century" in *Patients and Practitioners: Lay perceptions of medicine in pre-industrial society*, ed. Roy Porter (Cambridge, Cambridge University Press, 1985), 254. Both Andrew Wear and Ken Albala are more pessimistic about the effect these regimens had on non-medical people. Andrew Wear, *Knowledge & Practice in English Medicine, 1550–1680* (Cambridge, Cambridge University Press, 2000), esp. ch. 4; Ken Albala, *Eating Right in the Renaissance* (California, University of California Press, 2002), 2.

[26] L. J. Rather, "The 'Six Things Non-Natural': A Note on the Origins and Fate of a Doctrine and a Phrase," *Clio Medica*, 3 (1968).

[27] Saul Jarcho, "Galen's Six Non-Naturals: A Bibliographic Note and Translation," *Bulletin of the History of Medicine*, 44, 4 (July-August 1970).

despite the gradual disappearance of humoural theory.[28] Research has also focussed on how early modern texts about the six non-naturals changed over time, and historians have not quite agreed on whether this genre is best characterised by tradition or innovation. Jack Berryman, one of the first to investigate this topic, argued that the course of the non-natural tradition was mostly one of continuity from Antiquity until c. 1900.[29] Others, such as Maria Pia Donato, have since argued that at least in Italy, traditional health literature based on the non-naturals collapsed in the seventeenth and eighteenth centuries.[30]

In their recent book on conserving health in early modern Italy and England, Cavallo and Storey argue that, at least before the second half of the seventeenth century, the balance of all non-naturals stood at the heart of the health advice and that most regimens devoted a chapter or section to each of the non-naturals. However, after 1630, many popular medical texts in England started to focus exclusively on dietary advice and can no longer be considered traditional regimens, the authors argue.[31] Other historians such as David Gentilcore and Ken Albala have also noted a significant shift in this period, although they do not consider the later texts to be completely distinct from those that preceded them. Gentilcore argues that the 'regimen' returns after a short period of decreased popularity, albeit in a slightly different form: one which employs a 'new medical language' but otherwise remains true to the original structure of the regimen.[32] The texts became shorter, more generalised, and some non-naturals were discussed more briefly or left out altogether. In Cavallo and Storey's volume,

[28] Peter H. Niebyl, "The Non-Naturals," *Bulletin for the History of Medicine*, 45, 5 (1971).

[29] Jack Berryman, "The Tradition of the Six Things Non-Natural: Exercise and Medicine from Hippocrates through Ante-Bellum America", *Exercise and Sport Sciences Reviews*, 17, 1 (January 1989).

[30] Maria Pia Donato, "The afterlife of the non-naturals in early eighteenth-century Hippocratism: from the healthy individual to the healthy population," in *Conserving Health in Early Modern Europe: Bodies and environments in Italy and England*, eds. Sandra Cavallo and Tessa Storey (Manchester, Manchester University Press, 2017), 158–159; Sandra Cavallo, "Conserving health: the Non-Naturals in early modern culture and society", in *Conserving Health in Early Modern Europe: Bodies and environments in Italy and England*, eds. Sandra Cavallo and Tessa Storey (Manchester, Manchester University Press, 2017), esp ch. 1.

[31] Cavallo and Storey, (eds.), *Conserving Health in Early Modern Europe: Bodies and environments in Italy and England* (Manchester, 2017), ch. 1.

[32] David Gentilcore, *Food and Health in Early Modern Europe: Diet, Medicine and Society, 1450–1800* (London, Bloomsbury, 2016), 30–36.

Maria Pia Donato argues that medical writers were increasingly concerned only with specific topics such as the impact of air, food, and water on bodies.[33] Andrew Wear noted a similar trend in England and North America in that while overall Galenic and Aristotelian non-natural theories were decreasing in popularity, the non-natural 'air' grew in popularity.[34]

As eighteenth-century medical texts increasingly advised on some non-naturals rather than necessarily all, historians have similarly increasingly focussed their research on specific non-naturals rather than the doctrine of non-naturals as a whole. This trend of focussing on individual non-naturals or health-related topics may be representative of shifts in interest in the field of history more generally. Topics of everyday life became popular in late twentieth-century historical studies, including in the field of medical history. Topics including food and diet, personal and public hygiene, and sports were pursued in an effort to construct an image of the lives of 'ordinary people'. These more focussed approaches have led to works that help place the practice and theory surrounding the non-naturals in their wider context, an example of which is Sasha Handley's work on early modern sleep.[35] There has been some disparity in the topics which have been covered by historians, however, with the history of food and diet being significantly over-represented when compared to for example exercise—a phenomenon that Cavallo and Storey have since addressed and sought to rebalance.

Historical research into food and diet in the past has been significant compared to that of other non-naturals such as exercise, and it has diverged into many different directions. Much work has been done on shifts in medical theories about diet, especially in relation to the effects of chemical, vitalist, and, to a lesser extent, iatromechanical theories on dietetics.[36]

[33] Donato, "The afterlife of the non-naturals in early eighteenth-century Hippocratism," 158–159.

[34] Andrew Wear, "Place, Health, and Disease: The Airs, Waters, Places Tradition in Early Modern England and North America," *Journal for Medieval and Early Modern Studies*, 38, 3 (Fall 2008).

[35] Sasha Handley, *Sleep in Early Modern England* (New Haven, Yale University Press, 2016).

[36] Antonio Clericuzio, "Chemical and mechanical theories of digestion in early modern medicine", *Studies in History and Philosophy of Biological and Biomedical Sciences*, 43, (2012); Barbara Orland, "The fluid mechanics of nutrition: Herman Boerhaave's synthesis of seventeenth-century circulation physiology," *Studies in History and Philosophy of Biological and Biomedical Sciences*, 43, (2012); Elizabeth A. Williams, 'Sciences of appetite in the Enlightenment, 1750–1800', *Studies in History and Philosophy of Biological and Biomedical Sciences*, 43, (2012).

As early as the 1960s, Owsei Temkin assessed the impact on seventeenth- and eighteenth-century challenges to the Galenic system and argued that dietetic medicine had changed drastically by the early eighteenth century.[37] Temkin pointed out that Swiss physician and alchemist (c. 1493–1541) Paracelsus shifted the focus from diet to drugs, and Sydenham concerned himself with diseases rather than dietary errors, which, Temkin argued, ensured that food and diet no longer demanded the same level of attention in eighteenth-century medicine as they did before. The argument that the topic of diet lost popularity among physicians from the seventeenth century onwards has been repeated by others, such as Heikki Mikkeli, who similarly argued that dietetics was increasingly relegated to the fringes of medical research, in favour of physiology, pathology, and therapeutics.[38] Despite these conclusions that dietetics lost importance in this period, the many health guides published in the long eighteenth century indicate clearly that matters of food and diet continued to matter to physicians, as well as to non-medical people.

Others have assessed not only whether but exactly how dietetics were affected by new schools of thought. Elizabeth E. Williams argued that late eighteenth-century physio-chemical theories and experiments enabled physicians to establish universal standards of health which largely ignored individual appetites and needs.[39] Physicians started to give increasing credence to experimental procedures which claimed to offer certainties about the digestive process and the chemical makeup of foodstuffs. As a result, attention to individual constitutions, which had been a primary aspect of dietetics since Hippocrates, started to wane, Williams argued.[40] Very recently, Emma Spary has similarly stated that the emergence of food chemistry constituted one of the most salient transformations in early modern discussions of food, pointing especially at the invention of food groups and eventually the calorie.[41] She too found that individual factors disappeared from the dietetic discourse in favour of general rules.[42] Williams and Spary's research primarily focussed on the late eighteenth

[37] Temkin, "Nutrition from classical antiquity to the baroque".

[38] H. Mikkeli, *Hygiene in the early modern medical tradition* (Helsinki, Finnish Academy of Science and Letters, 1999).

[39] Elizabeth A. Williams, "Sciences of appetite in the Enlightenment, 1750–1800," *Studies in History and Philosophy of Biological and Biomedical Sciences*, 43, (2012).

[40] Ibid., esp. 393.

[41] Spary, "Food", 29.

[42] Ibid., esp 43–47.

and nineteenth centuries, and this book will further investigate these findings and assess to what extent they apply to the eighteenth century as a whole. Though much work has been done on the effects of medical theory on ideas about food and the body, there is still ample room for further enquiry. The effects of certain medical doctrines, such as that of iatromechanism, have not yet been fully investigated, leaving some aspects of eighteenth-century dietary advice unexplained. More importantly, however, existing scholarship on these doctrines tends to focus on the origin and longevity of the theories themselves. Questions about how such theories affected popular dietary advice remain largely unasked and unanswered. This book seeks to fill this gap in our knowledge by examining especially how iatromechanical ideas shaped the dietary advice of this period.

Other historians, notably Sandra Sherman and Rebecca Earle, have approached the topic of food and diets from a more political and economic angle, by investigating how authorities' cost considerations affected dietary advice dispensed to the labouring poor, especially in controlled environments such as work houses and hospitals.[43] As Spary notes, historians have long recognised and assessed the relationship between food and governance and there is ample scholarship seeking to correlate food intake with shifts in governance and economic growth.[44] These studies have contributed much to our understanding of changes in eating habits and dietary recommendations. Rebecca Earle has charted eighteenth-century efforts to define the nature of nourishment and argued that, in the eighteenth century, food was already considered central to the strength of

[43] Sherman, *Imagining Poverty*; Spary, "Food"; Rebecca Earle, "The Political Economy of Nutrition in the Eighteenth century," *Past and Present*, 242, 1 (Feb. 2019).

[44] Craig Muldrew, *Food, Energy and the Creation of Industriousness: Work and Material Culture in Agrarian England, 1550–1780* (Cambridge, Cambridge University Press, 2011). Steven L. Kaplan, *Bread, Politics and Political Economy in the Reign of Louis XV*, 2 vols. (Dordrecht, Springer, 1976); Alan Booth, "Food Riots in the North-West of England, 1790–1801", *Past and Present*, 77 (Nov. 1977); Carole Shammas, "The Eighteenth-Century English Diet and Economic Change", *Explorations in Economic History*, 21, 3 (July 1984); John Bohstedt, *The Politics of Provisions: Food Riots, Moral Economy, and Market Transition in England, c.1550–1850* (Ashgate, Farnham, 2010); E. C. Spary, *Eating the Enlightenment: Food and the Sciences in Paris, 1670–1760* (Chicago, Chicago University Press, 2012); David Meredith and Deborah Oxley, "Food and Fodder: Feeding England, 1700–1900," *Past and Present*, 222 (Feb. 2014); Bernard Harris, Roderick Floud and Sok Chul Hong, "How Many Calories? Food Availability in England and Wales in the Eighteenth and Nineteenth Centuries" *Research in Economic History*, 31 (Emerald Group Publishing Limited, 2015).

nations. Theorists of political economy, Earle explains, knew that a healthy nation needed healthy workers, and that healthy workers needed nourishing foods. These theorists set about determining how this workforce could be supplied with the required nourishment, preferably at a low cost, Earle found, and these motivations go a long way towards explaining late eighteenth-century scientific efforts to determine which foods were most 'nourishing'.[45] In *Imagining Poverty*, Sherman found that statistics and tables replaced considerations for the personal tastes and individual constitutions of the labouring poor when it came to determining their dietary needs and argued that this shift was motivated by economic considerations.[46] Investigations such as Earle's and Sherman's have significantly advanced our understanding and appreciation of the various contexts that we must consider as we research eating habits of the past, but many aspects about eighteenth-century dietary advice remain unclear. The investigations into the diets of the labouring classes ultimately tell us little about how and why the authors of health guides and similar literature determined what was and was not an appropriate aliment for the non-labouring members of society. This book will approach the question from a medical rather than a political/economic perspective and suggest that the cause for shifts in dietary advice aimed at the middling and upper classes may to be located in these medical theories as well as political or economic factors.

A number of studies have focussed on the dietary practices of non-medical people rather than the medical theories which underpinned them. Central to these studies is the question of the extent to which medical advice and dietary rules set by physicians reached the non-medical public and whether advice was followed or ignored. These studies often focus on matters of authority, especially on how this authority was established.[47] Steven Shapin, for example, has published several articles considering medical authority in relation to dietary advice. His 'Was Luigi Cornaro a Dietary Expert' argues that famous dieter Luigi Cornaro managed to constitute himself as a dietary expert because he lived in a historically specific

[45] Earle, "The Political Economy of Nutrition in the Eighteenth century", 84–85.

[46] Sherman, *Imagining Poverty*.

[47] Steven Shapin, "Trusting George Cheyne: Scientific expertise, common sense and moral authority in early eighteenth-century dietetic medicine," Bulletin of the History of Medicine, 77 (2003); Lucia Dacome "Living with the chair: private excreta, collective health and medical authority in the eighteenth century," History of Science, 39 (2001).

medical culture where personal experience counted as reliable evidence.[48] As will be discussed in this book, the eighteenth century witnessed several attempts by physicians to undertake dietary experiments on their own bodies. This phenomenon has already been documented by Lucia Dacome, whose 'Living with the Chair' investigated the extent to which such self-experiments served as evidence for the general functioning of bodies in a broad sense.[49] Studies such as these demonstrate the complicated nature of medical authority, especially in cases where dietary rules were based on experiments done on the author's own body. Despite the extensive work that has already been done in this area, the question of whether readers of medical literature accepted health rules based on other people's bodies and applied these to their own remains unanswered.

Research done on early modern health does not necessarily reflect the importance ascribed by eighteenth-century people to these topics as much as our own interests in these topics. Modern concerns with obesity, for example, go a long way in explaining recent historians' interest in early modern literature about diet and fatness. Following modern-day concerns with obesity rates, historical interest in people's relationship with food has grown in the past few decades. One of the first to produce a history of diets was Hillel Schwartz, with his 1986 *Never Satisfied: A Cultural History of Diets, Fantasies and Fat*.[50] Schwartz argued that a history of fat is also a history of dieting since 'the two have always gone hand in hand'. This is an idea that this book carefully investigates. While Schwartz considered 'slimming' to be 'the modern expression of an industrial society', recent historians have investigated further back in time for evidence of eighteenth-century concerns with diet and fatness. In fact, evidence of concerns with corpulency, as well as weighing practices, has led to some historians, such as Ken Albala, to argue that the history of our modern weight-loss practices started in the eighteenth century.[51] Lucia Dacome argues along the same lines, explaining that eighteenth-century representations of corpulence helped create an ideal embodiment and created fears

[48] Steven Shapin, "Was Luigi Cornaro a Dietary Expert?" Journal of the History of Medicine and Allied Sciences, 73, 2 (March 5, 2018).

[49] Dacome, "Living with the chair".

[50] Hillel Schwartz, *Never Satisfied: A Cultural History of Diets, Fantasies and Fat* (New York, Anchor, 1986).

[51] Schwartz, *Never Satisfied*, 5; Ken Albala, "Weight Loss in the Age of Reason", in *Cultures of the Abdomen: Diet, Digestion, and Fat in the Modern World* eds. Christopher E. Forth and Ana Carden-Coyne (Basingstoke, Palgrave Macmillan, 2005).

about excessive consumption.[52] Hence, it is in this context that much work focusses on the history of eating to gauge how people in the past have managed their food intake in relation to their body management. These studies of fat bodies have so far largely left out any discussions about exercise as a manner of body management, an oversight this study seeks to rectify.

Several historians have researched how eighteenth-century society viewed the body, how certain bodies were stigmatised, or idealised, and why certain body aesthetics were celebrated and others viewed negatively. As David Turner especially has argued, beauty was a topic of much interest in seventeenth- and eighteenth-century society, and body image a recurring theme in many periodicals, ballads, chapbooks, and other popular texts.[53] Turner's 'The Body Beautiful' investigates what defined a body as 'beautiful' and how beauty was viewed as a source of anxiety, as well as power, and the social value of personal appearance. As Turner found, seventeenth- and eighteenth-century society placed great importance on symmetry and proportion, and there was a sense that human beauty could be measured objectively.[54] In this book, I similarly argue that eighteenth-century physicians attached great importance to symmetry and proportion but add that physicians of this period were primarily concerned with the symmetry and proportions of the internal, rather than external body. More recently, historians such as Alun Withey and Solle Ylivuori have started to explore the role of bodily appearance in eighteenth-century British culture of 'politeness'.[55] As their work has shown, eighteenth-century men and women were increasingly expected to manage the appearance of their bodies, including their body shape.[56] Withey has argued that while previously body shape was seen as determined by God, this changed over the course of the eighteenth century as a new culture emerged which

[52] Lucia Dacome, "'Useless and Pernicious Matter': Corpulence in Eighteenth-century England," in *Cultures of the Abdomen: Diet, Digestion, and Fat in the Modern World*, eds. Christopher E. Forth and Ana Carden-Coyne (Basingstoke, Palgrave Macmillan, 2005), 193–199.

[53] David M. Turner, "The Body Beautiful", in *A Cultural History of the Human Body in the Enlightenment* ed. Carole Reeves (Oxford, Berg, 2010), 113.

[54] Ibid., 114.

[55] Alun Withey, *Technology, self-fashioning and politeness in eighteenth-century Britain: refined bodies*. Palgrave Macmillan, London, 2015; Solle Ylivuori, *Women and Politeness in Eighteenth-Century England: Bodies, Identities and Power*. Routledge, London, 2018.

[56] Withey, *Technology, self-fashioning and politeness in eighteenth-century Britain*, 1; Ylivuori, *Women and Politeness in Eighteenth-Century England*, 15.

emphasised the need to conquer the body.[57] The concept of eighteenth-century 'politeness' has been used as an idiom for 'civil' 'genteel' and 'mannered', of which conversation, education and manners were part, but appearance and form were also increasingly important factors to appearing 'polite', Withey found.[58] This development in attitude was accompanied by new technologies in the form of steel apparatuses which made it possible to alter and manage the shape and posture of the body, allowing 'deformed bodies' to be corrected back to their 'natural' state.[59] These technologies proliferated especially after 1750, which roughly coincides with the increased availability of another mechanical device designed to manage the human body, namely the weighing machine.[60]

Naturally, historical investigations into attitudes towards bodies of the past have tried to gauge the extent to which bodies of certain shapes and sizes were stigmatised, as they are today. Historical debate on how fat bodies were viewed and what levels of fat were considered ugly or beautiful, healthy or unhealthy, is very much ongoing, not least because social and medical usage of terms such as 'fat', 'overweight', and 'corpulent', and practices such as food measuring or body weighing can be difficult to interpret. Turner, for example, argued that while a 'certain plumpness' was seen as desirable and even as a sign of health, 'the truly obese were viewed with contempt', and that weight loss was recommended to such individuals to regain beauty.[61] Freya Gowrley's investigation of discussion of fat bodies in eighteenth-century prints, however, has found that in this period images and stories of extremely fat bodies were not necessarily presented or viewed as negative, a conclusion which will be further supported by this books' findings.[62] This book furthers the ongoing debate though a close investigation of discussions of fat bodies and body weight in late seventeenth- and eighteenth-century medical texts. It is this book's contention that fat bodies were not necessarily portrayed negatively, and when they were, this was not based on aesthetic factors such as their size or fleshiness. As will be shown, fat bodies were medicalised based on their (in)ability to exercise and partake in day-to-day duties. Their size, shape, or weight

57 Withey, *Technology, self-fashioning and politeness in eighteenth-century Britain*, 1.

58 Ibid., 7.

59 Ibid., 131–132.

60 See Chaps. 3 and 6 of this book for extensive discussions about the development and spread of weighing machines in eighteenth-century Britain.

61 Turner, "The Body Beautiful", 122.

62 Freya Gowrley, "Size Matters", *History Today*, 71, 12 (December 2021), 45.

played no part in physicians' conclusions as to whether they were healthy or unhealthy.

Concerns with bodily aesthetics do not seem to be as prominent in health guides as we might expect. In eighteenth-century texts, the 'ideal body' was primarily measured and determined with reference to the well-functioning of the inner body. Physicians and other authors of early and mid-eighteenth-century guides were, like the rest of society in this period, deeply interested in proportions, symmetry, and equilibrium, but their focus was predominantly on the internal processes of the body. The proportions or relative symmetry of the external body rarely featured in the health guides that this book investigates. The same is true for letters exchanged between physicians and their patients. In the letters written to physician William Cullen, for example, patients did not usually express concern about changes with the aesthetics of their bodies but were intensely worried about alterations inside their bodies, especially in relation to digestion and excretion. This of course does not mean that physicians were unaffected by contemporary attitudes to beauty, politeness, and the body. However, the internal workings of the body continued to play a great role in what constituted an 'ideal' body, and it was the preoccupation with the minutiae of internal physiological processes that drove weighing practices, rather than concerns with body shape. As such, this book not only furthers our understanding of discussions and practices related to the management of the body, but it also offers a different perspective on the role of the external and internal body in eighteenth-century discussions about the healthy and ideal body by approaching this from a medical, rather than cultural perspective.

The history of exercise has received little attention to date. Historians looking at the history of exercise have largely done so with a view to tracing back the beginnings of our modern sports and sport cultures.[63] Allen Guttmann's 1978 book, *From Ritual to Record*, differentiated modern

[63] Such as: Allen Guttmann, *From Ritual to Record: The Nature of Modern Sports* (New York, Columbia University Press, 1978); Neil Tranter, *Sport, economy and society in Britain, 1750–1914* (Cambridge, Cambridge University Press, 1998); Dennis Brailsforth, "Sporting Days in Eighteenth Century England," *Journal of Sport History*, 9, 3 (Winter, 1982); Stefan Szymanski, "A Theory of the Evolution of Modern Sport," *Journal of Sports History*, 35, 1 (Spring 2008); Alan Tomlinson and Christopher Young, "Towards a New History of European Sport," *European Review*, 19 (2011); Wray Vamplew, "Industrialisation and Sport," in *The Oxford Handbook of Sports History*, eds. Robert Edelman and Wayne Wilson (Oxford, Oxford University Press, 2017).

from pre-modern sport.[64] Using a definition of sport as 'playful physical contest' characterised by 'regulation, specialisation, bureaucracy and the focus on setting records', Guttmann argued that pre-modern sporting activities did not possess these characteristics and therefore are not included in his history of sports.[65] Guttmann's focus on setting apart 'modern sport' from pre-modern sport had the consequence of steering historians to place a sharp divide between them, with the boundary firmly placed in the early nineteenth century, or rather with 'the industrial revolution'. Neil Tranter, for example, argues that sports in pre-industrial societies were 'largely devoid of institutional structures and written rules' and that 'sport, in its modern, organised form was 'truly an invention of the Victorian and Edwardian age'.[66] Jean-Francois Bourg and Jean-Jacques Gouguet argued much the same, claiming that 'modern sport was born in England at the time of the industrial revolution', and Raymond Thomas also stated that 'sport was born in nineteenth-century England'.[67] In this way, early modern exercise was separated from 'sports' history, both in definition and in timeframe.

This does not mean that early modern exercise practices have not been investigated at all, but rather that they have rarely been discussed in a history of sports context. There are a few exceptions, however. Historians such as John McClelland, James Williams, Mike Huggins, and Wolfgang Behringer have researched sports in the pre-modern era. Their studies have focussed on trying to prove that sports in the early modern period possessed the characteristics that Guttmann ascribed to 'modern' sports. Behringer, for example, argues that the early modern period should be seen as 'the formative period of modern sport', due to the high levels of institutionalisation and standardisation that sports in the period 1450–1850 underwent.[68] Mike Huggins too argued that the early modern period already saw 'growing standardisation and the slow emergence of professionalism'.[69] James Williams argued that, at least amongst the elite

[64] Guttmann, *From Ritual to Record,* Preface.

[65] Ibid., 16–55.

[66] Tranter, *Sport, economy and society in Britain,* 1 and 16.

[67] Jean-Francois Bourg and Jean-Jacques Gouguet (2005), and Raymond Thomas (1991), both translated by Stefan Szymanski and published in "A Theory of the Evolution of Modern Sport," *Journal of Sport History,* 35, 1 (2008), 3.

[68] Wolfgang Behringer, "Arena and Pall Mall: Sport in the Early Modern Period," *German History,* 27, 3 (July 2009).

[69] Mike Huggins, "Early Modern Sport," in *The Oxford Handbook of Sports History,* eds. Robert Edelman and Wayne Wilson (Oxford, Oxford University Press (2017), 113.

members of society, sport became increasingly organised over the course of the sixteenth century.[70] McClelland pointed to physical education from the fourteenth century onwards.[71] Two years later, McClelland and Brian Merrilees edited a volume which explored pre-modern sport over several centuries, in which they argued that the sporting activities practiced in these centuries should be called 'sports' too, because pre-modern sports displayed organisation, structure, rules, and professionalism—many of the characteristics of modern sport, they point out.[72] As such, historians of early modern sports have challenged Guttman's exclusion of pre-modern sports in his definition of sport but not his definition of sport itself.

Not until 2007 was the use of this concept of 'modern sport' challenged by Martin Johnes, who exhorted historians of sport to question whether there even was a singular concept called sport and whether these activities we now call sports were considered collectively as connected kinds of physical activity in the pre-modern period.[73] In order to understand the wider implications of sports on societies, Johnes argues, historians need to make better connections between sports and other leisure activities, and need to study sport in history rather than the history of a certain sport or a limited definition of sport.[74] Tony Collins's article 'Work, Rest and Play: Recent Trends in the History of Sport and Leisure' also called for a broadening of perspective by proposing to study sport in relation to the field of history of leisure.[75] The history of sport, Collins argues, needs to be studied for what it can tell us about broader society, rather than for its own sake.[76] Paul Ward repeated this sentiment in 2013, calling for historians of sport to no longer 'ghettoise' sport history in separate chapters or sport history journals, but instead to broaden the appeal of their subject to mainstream historians and seek to publish their work in more mainstream history journals to integrate the history of sports into

[70] James Williams, 'Sport and the Elite in Early Modern England', *Sport in History*, Vol. 28, No. 3 (Sept. 2008).

[71] John McClelland, *Body and Mind: Sport in Europe from the Roman Empire to the Renaissance* (Abingdon, Routledge, 2007).

[72] John McClelland and Brian Merrilees (eds.), *Sport and Culture in Early Modern Europe* (Toronto, Centre for Reformation and Renaissance Studies, 2009), p. 22 and p. 36.

[73] Martin Johnes, "British Sports History: The Present and the Future," *Journal of Sports History*, 35, 1 (2008), 68–69.

[74] Ibid., 69.

[75] Tony Collins, "Review Article: Work Rest and Play: Recent Trends in the History of Sport and Leisure," *Journal of Contemporary History*, 42, 2 (2007).

[76] Ibid., 399.

wider histories.[77] This broadening of the field must also include a broadening of the definition of sport in sport historiography, especially to include other physical leisure activities.

In 2004 and 2008 Rebekka Mallinckrodt argued for the expansion of the 'sport paradigm' to cover the 'culture of movement' more broadly, but still centred her research around physical contests that had standardised rules, organisation, and other features that fitted Guttmann's definition of sports.[78] It is not until more recently that we find works which add new perspectives to the history of sport and exercise. Sandra Cavallo and Tessa Storey investigate changing definitions of what counted as 'exercise' between the late medieval period and the sixteenth century and how exercise advice was affected by shifts in social, cultural, and medical ideas and contexts.[79] Mallinckrodt and Angela Schattner's 2016 *Sport and Physical Exercise in Early Modern Culture: New Perspectives on the History of Sport and Motion* focusses on the early modern 'culture of movement' rather than the more restrictive 'sports' to include non-competitive forms of physical exercise. Their book contains several chapters focussing on health-motivated exercise, including one by Alessandro Arcangeli, in which he demonstrates the pressure put on renaissance women to be physically active because their moist bodies were thought to be in even more need of exercise that men's bodies.[80] In other chapters Tessa Storey and Sandra Cavallo trace exercise practices in family letters in sixteenth- and seventeenth-century Italy, while Martin Dinges's chapter looks at everyday movement in German-language ego-documents.[81] This

[77] Paul Ward, "Last man picked. Do mainstream historians needs to play with sports historians?" *International Journal of the History of Sports*, 30, 1 (Jan. 2013).

[78] Rebekka von Mallinckrodt, "Bewegte Geschichte – Pladoyer for eine verstarkte Integration und konzeptuelle Erweiterung der Sportgeschichte in der fruhneuzeilichen Geschichtswissenschaft," *Historische Anthropologie* 12, 1 (2004), 134–9; Rebekka von Mallinckrodt, *bewegtes leben. Korpertechniken in der fruhen neuzeit* (Wiesbaden, Harrassowitz, 2008), 1–14, esp 7.

[79] Sandra Cavallo and Tessa Storey, *Healthy Living in Renaissance Italy* (Oxford, Oxford University Press, 2013).

[80] Alessandro Archangeli, "Exercise for Women", in *Sports and Physical Exercise in Early Modern Culture*, eds. Rebekka Von Mallinckrodt and Angela Schattner (Abingdon: Routledge 2016).

[81] Sandra Cavallo and Tessa Storey, "'Decorous' and Pleasant Exercise: Competing Models and the practices of the Italian Nobility, in *Sports and Physical Exercise in Early Modern Culture*, eds. Rebekka Von Mallinckrodt and Angela Schattner (Abingdon: Routledge 2016); Martin Dinges, "Exercise, Health and Gender: Normative Discourses and Practices

volume thus marks the first attempt to link the field of sports history with the field of medical/health history and draws our attention to the role of the early modern culture of movement within wider cultures of health and medicine. With other chapters focussing on commercialisation and professionalisation, and status and identity, Mallinckrodt and Schattner have made not only a significant contribution to the field of sports and exercise history, but they have also changed and focussed its direction to have a much wider impact on early modern scholarship outside the hitherto limited field of sports history.

The expansion of the definition 'sports' to include a broader 'culture of movement' has led to Schattner and other historians making the link between sports and leisure that Johnes recommended. Schattner found that exercise facilities provided in places of leisure such as the early modern spa indicate to us that exercise was more closely entangled with other forms of leisure than previously thought.[82] Equally, commercialised sports venues offered a range of leisure facilities, she points out, and for contemporaries the differences between sport and/or exercise and leisure were not distinct.[83] Alessandro Archangeli and Steward Hill also found that exercise was entangled with the increasing interest in leisure, especially in eighteenth-century spa culture, where places dedicated to the health properties of water and air combined bathing, taking the air, and exercising.[84] As such, exercise was intrinsically connected with these other non-naturals to form part of the 'new' health culture. This has already been suggested by Ginnie Smith, who found that the hitherto much neglected theme of exercise was in fact much discussed in the medical literature post 1700 and that this was linked to the popularity of the 'cool regimen' and the emergency of balneology.[85] She suggests that the field of medical history is in

in Eighteenth- and Nineteenth-Century German-Speaking Countries", in *Sports and Physical Exercise in Early Modern Culture*, eds. Rebekka Von Mallinckrodt and Angela Schattner (Abingdon: Routledge 2016).

[82] Angela Schattner, "For the Recreation of Gentlemen and Other Fit Persons of the Better Sort: Tennis Courts and Bowling Greens as Early Leisure Venues in Sixteenth- to Eighteenth-Century London and Bath," *Sport in History*, 34, 2 (Jun. 2014).

[83] Ibid., 210.

[84] Ibid., 207–209; Alessandro Archangeli, *Recreation in the Renaissance: Attitudes towards Leisure and Pastimes in European Culture, c. 1425–1675* (Basingstoke, Palgrave Macmillan, 2003), esp. his chapter "medical discourse"; Steward Hill, *Leisure cultures in urban Europe, c. 1700–1870* (Manchester, Manchester University Press, 2016), 239–241.

[85] Smith, "Prescribing the rules of health", 259.

need of further research into sport and exercise.[86] Despite Smith's recommendations, subsequent research has not given exercise the place it deserves in any analysis of eighteenth-century health cultures.

In this book I investigate a wide selection of eighteenth-century British sources, which include primarily medical texts, but also letters, diary extracts, magazines, pamphlets, and ledgers. The focus on health guides has been determined by the degree to which it informs us of the intricate connection between medical theories and health advice. Health guides are here defined as medical texts which contain rules or advice as to how readers ought to conduct themselves in relation to their health. This definition excludes medical texts which do not contain any advice, but this book does use such texts in order to explore more in-depth the theories which informed advice. This examination of medical literature is supplemented with other material, both medical and non-medical.

Historians have widely debated how we might best research early modern ideas and practices related to health and the body. Source selection has been a topic central to this debate. Caveats include representation, with most sources reflecting only the opinions and practices of the most elite members of society, it is argued. Many sources only reflect the ideas and ideals of medical professionals but give us little to no insight into what non-medical people did in order to remain healthy. This book deepens and broadens our understanding of the relationship between physicians and non-medical people concerned with their health by gauging how physicians constructed advice with consideration for the readers of these texts. Such a focus allows us insight into the thinking process of physicians when considering how their theories could be translated into practicable actions. In doing so, this book does not limit itself to only physicians' patients, the people they treated directly, but also includes those who may have been affected by physicians' theories through published health advice. Health advice guides give us a glimpse of not only what physicians believed to be popular concerns, but they also allow us insight into how physicians attempted to make their advice palatable and practicable to their readers. These adjustments made by physicians consequently inform us of what physicians thought their readers believed to be reasonable or unreasonable health practices. This book argues that physicians wrote advice with their readers' willingness to partake in certain practices firmly in mind and that

[86] Ibid., 281.

health guides therefore should be carefully considered in any enquiry into health practices, alongside other sources such as letters and diaries.

This book presents a qualitative analysis of around 80 printed medical texts, most of which were published between 1680 and 1780, but with some pre-dating 1680 and some post-dating 1780. Texts from before 1680 serve to compare traditional advice to the newer, eighteenth-century advice, while texts from post-1780 primarily serve as a method of gauging how long certain trends or developments in health advice continued or started to wane. Though diet, exercise, and to a lesser extent, body weight, were discussed primarily in health literature, these discussions sometimes took place in wholly different kinds of sources, where the focus was not on supplying readers with health advice but rather on providing entertainment, for example.[87] As such, the analysis of discussions of body weight required an approach which utilised term-searches on various forms of the terms 'weight' and 'weighing', as well as 'corpulency' and 'fat/fatness' in online databases such as *Early English Books Online* (EEBO) and *Eighteenth Century Collections Online* (ECCO). In the case of 'weight/weighing', these terms yielded many results unrelated to medicine and the body and therefore required a limitation to sources which featured the term 'health' or 'non-naturals' in the title or key terms. The terms 'corpulency' and 'fat/fatness' were more limited in their results and did not require such limitations. Consequently, the searches for discussions about corpulency and fatness resulted in some finds outside the health guides, such as treatises reporting on the results of weighing experiments, but also in magazines and journals such as *Medical Transactions* and *The Spectator*.

The format of medical literature shifted over the period. Consequently, it can be difficult to divide these texts into neat categories, although some aspects are clearly discernible. Health guides generally had the term health (and/or long life) in the title and discussed a selection of non-naturals and contain advice about how to maintain and obtain health. The advice tends to be addressed to a wide audience, and they differ in this respect from medical texts which contain advice for a particular subset of people, such as the aged, young children, nursing mothers, or people suffering from a particular disease such as corpulency or gout. In these texts, advice is

[87] Such as the account of the 'extraordinary' Daniel Lambert: Anon., *The Life of that wonderful and extraordinarily heavy man, Daniel Lambert, from his birth to the moment of his dissolution; with an account of men noted for their corpulency, and other interesting matter* (New York: Samuel Wood & Sons,1818); https://wellcomecollection.org/works/xycvjyp8.

constructed around what physicians perceive to be physiological aspects unique to the old, the young, nursing mothers, or the sick, and this advice was understood to be inappropriate for anyone not belonging to those particular groups.[88] General health guides, however, were designed for a much wider audience, even if there is ample evidence to indicate that physicians believed their readership to have certain characteristics.[89] The general health guide covered a wide range of health-related topics which generally included each or most of the traditional six non-naturals (air, diet, exercise, excretions, sleep, and the passions of the mind), and sometimes other topics related to health which did not fit into those six sections. These guides contained very broad and general advice, dedicating a similar amount of space to each topic, although there were some discrepancies here.[90] They differed from medical treatises which focussed on a particular topic and which started to appear with increasing frequency from the seventeenth century onwards.[91] These latter texts generally treated one particular topic in-depth and their purpose was to make a case for the importance of that particular topic rather than to provide general guidance for living healthily or prolonging life. Health guides thus form a particular source base within the broader genre of medical literature, and it is on these texts that this book will focus in particular. From the selection of 80 medical texts, 32 fit into this category.

The other texts consist of medical treatises which focus on a particular illness, on a specific non-natural, or on medically distinct groups of people, and therefore require a slightly different approach to the more general health guides. Health advice in these texts tended to be more specific and detailed as they were constructed with a very particular group of people in

[88] They were believed to have unique constitutions, for example, or particular dietary or exercise needs due to their particular circumstances.

[89] As the discussions in the following chapters will demonstrate, physicians appeared to address a readership they believed to be predominantly male, above the age of 40, sedentary, and belonging to the middling to upper class.

[90] Diet was usually the largest topic, though the gap grew smaller over the course of the eighteenth century.

[91] Stolberg argued that the uptick in treatise focussed on particular medical topics and diseases may have been due to university-trained physicians needing to publish a dissertation on a particular topic, which consequently ensured a growing number of medical publications on virtually every topic. More particular to the eighteenth century, Stolberg pointed out that those physicians who advocated a new medical system, like Iatromechanists, could use particular diseases to offer more compelling explanations than their competitors. In Stolberg, "Abhorreas Pinguidem", 374.

mind. The rules found in these texts do not represent what physicians might have recommended to other groups of people, and so the advice herein needs to be considered carefully with the appropriate medical context in mind. These sources lend themselves particularly well to an examination of how rules were constructed differently for the different sections of society, and how physicians grouped patients together during a period in which the humoural classification of bodies started gradually to lose favour. This book will utilise such sources especially for examining which classifications physicians started to use instead of humoural differences. In doing so, it will engage with arguments presented by Michael Stolberg, whose research has indicated that already from the seventeenth century onwards, physicians took an ontological, rather than humoural approach to prescribing medication. This book gauges the extent to which this applied to dietary and exercise advice as well, for which purpose these sources lend themselves particularly well.

Historians have questioned and debated who read these texts and to what extent readers applied the advice contained within them to their day-to-day lives. William Coleman attempted to specify the socio-economic limits of health advice by arguing that advice on managing the non-naturals was above all aimed at elite bourgeois individuals to apply to themselves.[92] Bryan Turner in 1985, and later Andrew Wear in 2000, reiterated this argument and concluded that there is almost no way to tell from the literature whether there was a 'trickle down' of this 'elite' literature to the population at large and that therefore the effect of non-natural theory on lay practice is difficult to measure.[93] Recent work by Ken Albala ended on the same note, concluding that while advice on the non-naturals was embraced by the elite as an 'entertaining topic of conversation', questions as to whether the medical advice was adhered to, and by whom, remain unanswered.[94] Other historians who have tried to answer such questions include Ginnie Smith, Antoinette Emch-Deriaz, Michael Stolberg, and, most recently, Sandra Cavallo and Tessa Storey.[95] Smith

[92] William Coleman, "Health and Hygiene in the 'Encyclopedie': A Medical Doctrine for the Bourgeoisie," *Journal of the History of Medicine and Allied Sciences*, 29, 4 (Oct. 1974).

[93] Bryan Turner, "The government of the body: medical regiments and the rationalisation of diet", *British Journal of Sociology* 33 (1982); Wear, *Knowledge & Practice in English Medicine*, 165.

[94] Albala, *Eating Right in the Renaissance*, 2 and 171.

[95] Smith "Prescribing the rules of health": Antoinette Emch-Deriaz, "The non-naturals made easy," in *The Popularization of Medicine, 1650–1850*, ed. Roy Porter (Abingdon,

argued that the regimen's general popularity and longevity would have left few sections of the population untouched by advice about the non-naturals, and Emch-Deriaz pointed out that many authors and printers of these regimes explicitly stated their intention to make humoural advice available to a broader social stratum.[96] Such statements tell us little about actual readership, however.

This book will engage in this historiographical debate by following the methodological approach that recent historians such as Cavallo and Storey have used. I will combine the assessment of patient documents such as letters and diaries with printed medical literature to ascertain to what extent lay persons were aware of medical theories concerning the non-naturals, to what extent they understood such theories, and to what extent they applied the advice given in regimens to their day-to-day lives. Manuscript sources have increasingly been considered by historians focussing on dietetics. Michael Stolberg, Steven Shapin, Sandra Cavallo, and Tessa Storey have explored both letters and casebooks to investigate the relationship between medical theory and dietary practices.[97] Stolberg has warned that, like health guides, eighteenth-century letters reflect largely the upper and middling classes of society and that the medical attitudes of the lower classes are much more difficult to gauge.[98] Much of what we know from ordinary people comes, in fact, from accounts by educated contemporaries: physicians' case histories and letter consultations. These can be rich sources for researching the wider social strata, but first-person accounts of persons from the lower classes are rare.

Historians have attempted to find 'the patient's view' on health and health management through manuscript sources such as letters, diaries and other 'ego documents'.[99] A number of different findings have been

Routledge, 1992); Michael Stolberg, *Experiencing Illness and the Sick Body in Early Modern Europe* (London, Palgrave Macmillan, 2011); Sandra Cavallo and Tessa Storey, *Conserving Health in Early Modern Europe: Bodies and environments in Italy and England* (Manchester, Manchester University Press, 2017).

[96] Smith, "Prescribing the rules of health," 254; Emch-Deriaz, "The non-naturals made easy" 136–137.

[97] Shapin, "Trusting George Cheyne"; Sandra Cavallo and Tessa Storey, *Healthy Living in Late Renaissance Italy* (Oxford, Oxford University Press, 2013).

[98] Stolberg, *Experiencing Illness,* 14.

[99] One of the first to do so was Joan Lane, "The doctor scolds me': The diaries and correspondence of patients in eighteenth-century England," *Patients and practitioners: Lay perceptions of medicine in pre-industrial society* ed. Roy Porter (Cambridge, Cambridge University Press, 1985).

disseminated following research into such ego-documents, mostly supporting the hypothesis proposed in the aforementioned work by Smith.[100] Wayne Wild surmised from his reading of eighteenth-century British letters exchanged between patients and practitioners that lay persons were well-versed in medical matters and able to keep up with medical knowledge and jargon, suggesting that patients did indeed read a range of medical texts.[101] Roy Porter and later Alisha Rankin also found that patients were familiar with medical theory and added that lay persons used their medical knowledge to influence the direction their physician would take in healing them.[102] Barbara Duden argued along the same lines, showing that in letters, patients often expressed therapeutic preferences, displayed knowledge of a range of treatments, and argued with their physician about which treatments might suit their bodies and lifestyles best.[103] Michael Stolberg's reading of German and French ego-documents has led him to conclude that lay people expressed significant interest and concern in their own health.[104] While, as Micheline Lious-Courvoisier and Severine Pilloud have warned, letters 'do not lead to an objective truth about patients', they do give us access to lay perceptions, knowledge, and attitudes that is hard to find in other sources.[105]

Despite these findings, Stolberg argues that the impact of the regimen genre on the everyday life of most early modern people should not be overestimated.[106] He argued that the management of the non-naturals became an important issue only when someone fell ill, which suggests that contrary to their purpose, regimens did not influence lay people's

[100] Smith, "Prescribing the rules of health:" 254.

[101] Wayne Wild, "Medicine-by-Post: The changing voice of illness in eighteenth-century British consultation letters and literature", *Clio Medica*, 79, 1 (2006), 1–286, 19.

[102] Roy Porter, *Patients and Practitioners: Lay perceptions of medicine in pre-industrial society* (Cambridge, Cambridge University Press 1985); Alisha Rankin, *Panaceia's Daughters: Noblewomen as healers in early modern Germany* (Chicago, Chicago University Press, 2013).

[103] Barbara Duden, *The Woman Beneath the Skin: a doctor's patients in eighteenth-century Germany*, trans. Thomas Dunlap (Cambridge, Harvard University Press, 1991); Stolberg, *Experiencing Illness.*

[104] Stolberg, *Experiencing Illness.*

[105] Micheline Lious-Courvoisier and Severine Pilloud, "Consulting by letter in the eighteenth century: Mediating the Patients's View?" in *Cultural Approaches to the History of Medicine: Mediating Medicine in Early Modern and Modern Europe*, eds. Willem de Blecourt, and Cornelie Usborne (Basingstoke, Palgrave Macmillan, 2004), 72.

[106] Stolberg, *Experiencing Illness*, 44.

day-to-day lifestyles.[107] Of course, when well, people would have had little reason to write to their physician, and so we are left with not much written evidence of lay health management outside of periods of illness. Absence of evidence should not be taken to mean that healthy people were not concerned with maintaining their state of health, however, and further research into leisure activities, eating habits, and other potentially health-related customs could shed light on this. This book will contribute to this debate by its assessment of health guides, especially by looking at how physicians constructed advice with the reader and their lifestyle in mind.

This study seeks to address the wider question of the extent to which medical advice was read and followed by the non-medical urban elite, and to answer this I investigated groups of letters exchanged between physicians and their patients. These include letters by physician George Cheyne to his patient and friend Samuel Richardson, exchanged between 1733 and 1743 and published in a collection as *The Letters of Doctor George Cheyne to Samuel Richardson*.[108] This study has also utilised word and phrase searches on letter database *The Cullen Project*, which contains consultation letters to and from Scottish physician William Cullen (1710–1790).[109] These letters provide further insight into the sometimes complicated power dynamics between physician and patient in terms of setting the rules for and limitations of diet and exercise, something which has already been explored by Barbara Duden, whose work this book builds upon and extends.[110]

Historians of early modern health have also looked at diaries and autobiographies to learn about lay health practices. Joan Lane found diary entries discussing spa visitations and passages where the diarist complained about his/her diet, suggesting that lay persons engaged in more health-motivated activities whilst in a general state of health than might be expected.[111] Diary entries are therefore a potentially rich source of information regarding dietary and exercise practices; however, an exhaustive examination of diaries would require extensive and time-consuming search, the results of which could potentially prove fruitless. What is more,

[107] Ibid.

[108] *The Letters of Doctor George Cheyne to Samuel Richardson (1733–1743)*. Edited by Charles F. Mullett. Columbia: University of Missouri, 1943.

[109] "The Consultation letters of William Cullen", Royal College of Physicians at Edinburg. Accessible online via *The Cullen Project*, <http://www.cullenproject.ac.uk>.

[110] Duden, *The Woman Beneath the Skin*.

[111] Lane, "The doctor scolds me", 216.

one of the primary objectives of this book has been to ascertain the extent to which physicians constructed advice around their individual patients, which is something that non-medical texts such as diaries are unlikely to help with. Anecdotal non-medical evidence therefore serves this project in an auxiliary, rather than primary fashion, and an extensive literature review has yielded a number of diary sources relevant to this study. Examples from diary discussions about diet, exercise, or body weight located through the work of other historians are here considered in light of findings from the health guides, leading to new interpretations of familiar sources.

Finally, this book examines and re-evaluates a source discovered by Lucia Dacome: the weighing ledgers of an eighteenth-century grocery store, now a wine merchant located in central London.[112] These eight ledgers contain the weighing entries, among some other notes, of thousands of individuals who weighed their bodies in the store from 1765 until the early twentieth century. Although Dacome has made some excellent and interesting observations about these ledgers, much remains to be examined, and this book will contribute significantly to our understanding of the information that can be found within them. It will show how this information can be understood within the context of eighteenth-century ideas about body size and weight, diet, exercise and health.

Book Structure

Quantification and Measurement in British Health Advice c. 1670–1830 is split into six chapters informally organised in four sections. The first section consists of Chap. 1 and covers the introduction to the book. The second section, comprising Chaps. 2 and 3, explores eighteenth-century dietary advice, with Chap. 2 focussing on advice about the selection of aliment/food, while Chap. 3 looks at advice about dietary quantities.[113] Both chapters discuss the changes that dietary advice underwent in the eighteenth century, as well as those aspects of traditional advice which remained relevant. Combined, these chapters examine the dietary choices that physicians suggested their readers make, in terms of both what and

[112] A collection of eight 'weighing' ledgers is currently kept at 6 Berry Bros. and Rudd, a wine merchant on No. 3 St. James Street, London.

[113] The term 'aliment' comes from the Latin 'alere', meaning 'to eat' and can be found in eighteenth-century texts as a term for food. *Merriam-Webster.com Dictionary*, s.v. "aliment," accessed March 3, 2023, https://www.merriam-webster.com/dictionary/aliment.

how much readers ought to eat. The third section, comprising Chaps. 4 and 5, focusses on exercise advice. Chapter 4 explores the impact of iatromechanism on exercise advice as well as growing concerns with sedentarism, while Chap. 5 focusses on fears concerning sweat and perspiration and investigates how the iatromechanical school of thought contributed to these fears. Together these chapters explore how old and new theories came together to create historically singular exercise advice according to which readers had to move carefully to find the right balance between too much and too little exertion. Finally, the fourth section of the book consists of a single chapter which assesses the theory and practice of body weighing in this period. It combines considerations of the previous two sections and investigates how these are linked to the concept of body weight and how fears related to diet and exercise also found expression in the practice of body weighing.

Chapter 2 tracks changes in dietary advice in early eighteenth-century health advice following the growing popularity of the iatromechanical school of thought among physicians. The chapter explores how physicians imagined the body and its digestive processes to work and demonstrates that the impact of iatromechanical theory on health advice was significant even if it did not result in a complete overhaul of the entire genre. An assessment of the advice literature will show that some physicians departed from the traditional approach, which consisted of offering bespoke advice only applicable to the patient for whose individual constitution this advice was designed. Instead, these physicians attempted to construct general rules for large groups of people whom they considered physiologically similar enough to warrant generalisations. That dietary advice changed significantly somewhere between the eighteenth and the nineteenth century is well-documented, and historians have considered several possible factors which may have contributed to what is described as a shift from individual to general health advice, consisting mostly of economic and political considerations. These have so far provided a partial but not wholly satisfactory explanation of differences between early eighteenth-century physicians, whose advice had been primarily constructed for one patient alone and was therefore very singular in nature, and nineteenth-century physicians, whose rules of diet were of a more general nature and applied to large sections of society. This chapter seeks to address this gap through a careful assessment of eighteenth-century advice found in published medical literature which can be characterised as health guides. The advice given in these guides was generally addressed to the middling sorts and upper classes of society and therefore helps us establish if, firstly, this transition from the bespoke to the general also took place in the medical discussions relating to the well-to-do rather than just the working classes and, secondly, whether such a shift was motivated by similar or different ideas. Chapter 2 investigates these guides for indications that physicians shifted from advising individuals to instead providing dietary rules for large groups of people. It pays attention to advice related to the choice of food and assesses the extent to which individual factors such as constitution, age, illnesses, and living circumstances and personal taste were still considered by eighteenth-century physicians in the construction of the advice they printed in health guides.

Chapter 3 continues from Chap. 2 in its analysis of shifts in the dietary advice found in eighteenth-century health guides but focusses on advice relating to dietary quantities rather than qualities. It asks the question whether there was already an eighteenth-century shift where physicians assessed food quantities by measure, rather than the traditional method of following bodily cues such as appetite and fullness. This topic of how physicians of the past determined food quantities has featured extensively in recent historical discussion about dietary advice. Historians such as Emma Spary and Rebecca Earle have sought to explain the origins and motivations behind the quantitative approaches to diet that we generally see in early nineteenth- and late eighteenth-century literature, and this chapter directly contributes to this ongoing debate. Where this chapter covers new ground is by examining earlier sources to compare early eighteenth-century quantification methods with the later ones and gauging the extent to which how far these are connected. Both Spary and Earle suggest that the cause behind the late eighteenth-century shift towards quantifiable dietary measures was located in political and economic motivations, but I explore how medical theories and trends had a bearing on this development. This chapter consists of a detailed analysis of the measuring methods that were discussed in the medical literature of the eighteenth century and assesses to what extent these were recommended, and rejected, and how exactly these might have been applied in day-to-day life. It will look closely at how physicians discussed the benefits and the drawbacks of methods such as weighing food, counting spoonfuls, or measuring 'by eye', and how they arrived at their judgement as to whether such methods were sufficiently practical or overly cumbersome. Finally, the chapter also takes a closer look at the weighing experiments that took place over the course of the eighteenth century and explores how these experiments mirrored or departed from what was written in health guides. The answer to the question of whether changes in medical advice ultimately led to changes in medical practice can in part be answered by looking at the advice itself. The chapters will all look closely at how physicians arrived at the advice that they published and how considerations for the body, lifestyle, and circumstances of the reader featured in the advice given. Chapters 2 and 3 both assess the degree to which physicians accounted for personal tastes in discussions of how to select healthful food, as well as the embodied experiences of individuals in relation to feelings of appetite and fullness in the determining of healthy quantities.

Chapters 4 and 5 rectify the paucity of attention paid to the topic of exercise by historians of the early modern period, and the first of these chapters investigates increasing eighteenth-century concerns with sedentarism. Chapter 5 asks to what extent such concerns were fuelled by factors relating to urbanisation or new medical theories which emphasised the importance of exercise to the health of the body. It looks closely at the factors discussed by physicians in the medical literature, which included urbanisation, increased wealth, and changing leisure habits. The chapters also take into consideration physicians' medical explanations for their concern with what they perceived to be decreasing levels of exercise amongst their contemporaries. These chapters also investigate how exercise advice itself changed in the eighteenth century, especially in relation to new iatromechanical ideas of how exercise affected the body. It covers which exercises physicians tended to recommend in their advice, and at the medical grounds on which they based these recommendations. Furthermore, it also investigates how physicians navigated factors such as physical strength, access to exercise equipment, but also expendable income, leisure time, and even readers' willingness to partake in exercise, in determining the kind and measure of exercise they would recommend. The chapter therefore closely engages with one of the broader questions this book asks, namely how physicians constructed their advice to be practically applicable by a non-medical public.

Chapter 5 follows on from Chap. 4 in its focus on eighteenth-century exercise advice but concerns itself primarily with one particular aspect of this advice: the eighteenth-century argument that exercise was to remain sweat-free. It investigates the theories behind physicians' common saying that exercise should be 'ad ruborem, non sudorem' (to redness, not to sweating) and places these in the context of older, Santorian explanations of the body's balance between the *ingesta* and *excreta*. *Ingesta* were generally understood to constitute food and drink but could also include moisture absorbed through the skin (such as by bathing or from the air). *Excreta* constituted all matter expelled from the body through a variety of channels. These included stool and urine, but also perspiration, mucus, and blood. The chapter looks closely at physicians' understanding of the relationship between diet and exercise and how Santorian and iatromechanical understandings of this relationship affected exercise advice of this period. In line with the other chapters of this book, this chapter too consists of an investigation of the methods that people were advised to use to quantify exercise and measure perspiration. It therefore engages with the broader question of the extent to which we see a shift towards quantifiable research methods of measuring health and the body. Following on from Chap. 4, Chap. 5 gauges physicians' considerations of the practicalities of measuring perspiration. Eighteenth-century health advice continued to take into consideration individual factors such as constitution, age, and illnesses, as will be demonstrated throughout this book. These factors played a role in exercise advice and complicated seemingly straightforward rules such as that one ought to refrain from sweating. Similarly to Chap. 4, Chap. 5 will also feature a discussion of the extent to which the readers' personal sensations of fatigue continued to play a role in determining when one had exercised sufficiently, or too much. The final section of this chapter investigates how and why the discussion about sweat and exercise continued longer than some other aspects of Santorian and Iatromechanical health advice.

In the last chapter of the book, aspects from the previous chapters come together in a discussion of the nascent practice of body weighing, where concerns with the *ingesta* and *excreta* found expression. Although Santorio had first started to weigh himself in the sixteenth century, this practice became more widely practised in the mid-eighteenth century, both by medical professionals and by non-medical people. Chapter 6 gauges how eighteenth-century weighing practices compare to modern practices and situate them in their relevant medical context, especially that relating to the re-publication of Santorio's *Medicina statica* in the early eighteenth century. Also included is a discussion about corpulency and how body weight featured in literature about corpulency. The chapter asks to what extent these topics were discussed in tandem, in order to answer wider questions about the history of both body weighing and fatness. It also looks closely at how eighteenth-century physicians adjusted recommendations initially made by Santorio in the early seventeenth century in order to make body weighing more practicable by the non-medical public similarly to how they had adjusted dietary and exercise advice. Such investigations into physicians' decision-making in the construction of their advice help answer larger questions regarding the relationship between medical theory and popular reception of medical theory, and thus this book furthers discussion of how non-medical readers might have applied advice in their day-to-day lives.

From these chapters together, a number of conclusions can be drawn. Firstly, this book finds that, to an extent, health advice did become increasingly quantified over the course of the eighteenth century, which is in line with findings from other historians, especially in the field of the history of diet and body weight. Eighteenth-century health guides were concerned with finding the limits of healthful dietary and exercise quantities, and quantifiable methods for measuring aliment, excretions, and body weight were practised by medical, as well as non-medical individuals. This shift towards quantifiable methods for determining non-natural matters also extended to exercise advice replacing traditional advice. Where my book departs from previous findings, however, is in my assertion that eighteenth-century quantification methods did not yet result in standardised health advice. My research shows that while several physicians did shift advice to focus on groups with shared features rather than individual bodies, these physicians clearly continued to communicate to their readers that individual factors needed to be taken into account in any health-related decision making. The eighteenth century was a transitional period, during which the co-existing of older Galenic theories with new iatromechanical ones resulted in a medical milieu characterised by both continuity (in the persistence of attention to individual factors) and change (the grouping together of individuals with shared factors).

The research into physicians' health advice furthermore indicates that individual circumstances, lifestyles, and constitutions continued to inform the manner in which physicians constructed their advice until the late eighteenth century. The question of the relationship between medical literature and its assumed readership has long dominated historical debate, and this book argues that advice found in health guides was constructed with the reader in mind. This included considerations of whether non-medical readers would find certain dietary, exercise, or weighing practices used by medical professionals too cumbersome to apply in their day-to-day lives. My assessment of the weighing ledgers especially shows that there was a discrepancy between medical and non-medical weighing practices, which I argue is best explained by physicians' considerations of the practicalities of quantifiable measuring methods outside the medical profession, although letters and diary extracts suggest that the same applied to dietary and exercise practices.

Finally, this book demonstrates that the health advice of the eighteenth century was singular in nature and the product of a unique medical milieu. Many theories about health and models of the body co-existed in this period, resulting in health advice which was highly complex and contested. Direct comparison with modern dietary, exercise, and weighing practices risks obscuring those factors which shaped health advice and is therefore best avoided. Many historians have understandably attempted to locate the beginning of our modern concerns with diet, exercise, and body weight in health advice of the past by searching for the first indications of concerns with dietary quantities, weight-loss-motivated exercises, and unhealthy levels of fat. While these concerns can indeed all be found in eighteenth-century health advice, this book argues that none of these eighteenth-century discussions constitutes the beginning of our modern practices. Eighteenth-century discussions about diet, exercise, and body weight were motivated by specifically eighteenth-century health concerns, and a key aim of this book is to identify these.

CHAPTER 2

'Whats good and bad': Dietary Choices in Eighteenth-Century Health Guides

For an eighteenth-century physician, it was almost impossible to sit down at a meal without his company asking him many questions regarding the wholesomeness or unwholesomeness of particular ingredients, physician Francis de Valagin (1725–1805) noted.[1] That the choice of food was a popular topic in the seventeenth and eighteenth centuries has been echoed by historians who have investigated this period's numerous health publications. What is less certain, however, is how and why authors and readers of these publications made the food choices that they did. Much of the existing scholarship on early modern dietetics has focussed on the ebb and flow of medical schools of thought such as Galenism, iatromechanism, and iatrochemistry, or the extent to which medical advice was actually adopted by a non-medical population. Andrew Wear argued that despite the popularity of these guides, few readers took up the advice.[2] This, of course, does not mean that individuals did not make any alimentary choices—but Wear suggests that medicine's influence on dietary choices was limited. Later historians have contested Wear's pessimistic conclusions regarding the role

[1] Francis de Valagin, *A treatise on diet, or, the management of human life; by physicians called the six non-naturals* (London: Printed for the author by J. and W. Oliver, 1768), Eighteenth Century Collections Online, p. 94.

[2] Andrew Wear, *Knowledge & Practice in English Medicine, 1550–1680* (Cambridge: Cambridge University Press, 2000), 154.

M. Wübbels, *Quantification and Measurement in British Health Advice, c. 1670–1830*, Palgrave Studies in Medieval and Early Modern Medicine,
https://doi.org/10.1007/978-3-032-32760-4_2

of health guides in the dietary practices of the non-medical population. David Gentilcore has since argued that there was ample evidence of both the desire to learn and follow dietary advice, and so an enquiry into dietary advice of this period is a particularly appropriate starting point to investigate motivations for dietary choices.[3]

In this chapter I will show that the iatromechanical school of thought had a significant effect on dietary advice of the eighteenth century. Iatromechanical theories affected what qualities as well as what quantities of food were recommended. In this chapter, we will focus on the former—the qualities of food—while Chap. 3 will discuss the matter of quantity in dietary advice. In addition, this chapter will argue that there is evidence to suggest that during the eighteenth century, dietary advice became, at least to some extent, more generalised for certain groups of people, by which I mean that some physicians started to make dietary recommendations applicable to many people, rather than just one individual. Though bespoke dietary advice remained a feature of eighteenth-century medicine, we can perceive a shift towards dietary regimens aimed groups of people, particularly the old, the sedentary, and those generally considered 'weak' or 'sickly'. However, as this chapter will demonstrate, this shift did not apply evenly over eighteenth-century medical literature as a whole. It will argue that whether or not physicians prescribed rules for individuals or groups depended on the medical model that they used and what certain food was supposed to be good for. The more certain food types targeted what physicians considered to be shared features among their patients (such as a weak stomach), rather than individual features (such as temperaments or living circumstances), the more generally applicable the advice became. This chapter does not necessarily concern itself with the question of which food eighteenth-century physicians did or did not recommend, but rather, with their lines of reasoning for making such recommendations.

Some excellent research on the eighteenth-century efforts to determine the wholesomeness of food already exists and historians have offered a multitude of possible factors which may have affected how and why physicians were so keen to determine which foods were wholesome and which were not. Rebecca Earle has charted eighteenth-century efforts to define the nature of nourishment. Earle connects eighteenth-century discussions

[3] David Gentilcore, *Food and Health in Early Modern Europe: Diet, Medicine and Society, 1450–1800* (London, Bloomsbury, 2016), 5.

of nourishing food to concerns of a political and economic nature, and she has not been the first to do so.[4] Theorists of political economy, Earle explains, knew that a healthy nation needed healthy workers, and healthy workers needed nourishing foods. These theorists set about determining how this workforce could be supplied with the required nourishment, preferably at a low cost.[5] As this chapter looks at health guides and consequently the dieting habits of primarily the middling and elite members of society, the motivation of providing nourishment at minimum cost does not apply. The physicians who sought to determine which foods were nourishing were not interested in saving their clientele money on their groceries. Instead, they were walking a careful line where they, on the one hand, incited sufficient change to their patients' diet to cure their ills while also, on the other, avoiding upsetting the lifestyle of their patient too much. Significant changes and inconveniences to diet or lifestyle were unpopular among these physicians' patients and would likely cause the patient to not follow through on the physicians' advice.[6] The very nature of the readership of these guides may therefore go a long way in explaining why so many physicians continued to emphasise that dietary rules needed to be constructed around the body and lifestyle of the patient and not the other way around.

Nevertheless, these enquiries into the food requirements of the labouring classes may have inadvertently also affected the dietary advice of the health guides for the elites/higher classes by fostering a strict distinction between nourishing foods which were for labouring bodies and less nourishing foods which were suitable for those who did not labour. As other historians have demonstrated, class distinctions determined which foods were suitable for whom perhaps as much as medical theory or experiment did.[7] Ideas about masculinity and the nature of English people sometimes

[4] Earle, "The Political Economy of Nutrition in the Eighteenth century", 83. See also Barbara Orland and E.C. Spary, "Introduction" *Studies in History and Philosophy of Biological and Biomedical Sciences*, 43 (2012).

[5] Earle, "The Political Economy of Nutrition in the Eighteenth century", 84–85.

[6] Examples abound in physicians lamenting their patients' unwillingness to follow rules of health, but the best example is perhaps physician George Cheyne's letters to his friend Samuel Richardson, published in: *The Letters of Doctor George Cheyne to Samuel Richardson (1733–1743)*. Edited by Mullett, Charles, F., Columbia: University of Missouri, 1943.

[7] Kevin Siena, "Pliable Bodies: The Moral Biology of Heath and Disease", in *A Cultural History of the Human Body: In the Enlightenment*, ed. Carole Reeves (Oxford, Berg, 2010), 46–47; David Gentilcore, *Food and Health in Early Modern Europe: Diet, Medicine and Society, 1450–1800* (London, Bloomsbury, 2016), ch. 3.

played a role in determining the suitability of certain diets or aliments as well. Some opponents of the meat-free diet warned readers that vegetable food bred effeminate bodies and that the hardy English required stronger diets than people of other nations, for example.[8] The suitability of aliment was therefore determined by a number of cultural, economic, and political, as well as medical factors, all of which need to be kept in mind when attempting to gauge the reasoning behind physicians' advice as to which foods to consume and which to avoid.

Investigations such as Earle's have significantly advanced our understanding and appreciation of the various contexts we must consider as we research eating habits of the past, but many aspects about eighteenth-century diet advice remain unclear. Studies of the diets of the labouring classes ultimately tell us little about how and why the authors of health guides and similar literature determined what was and was not an appropriate aliment for the non-labouring members of society. This chapter seeks to address this imbalance by analysing early eighteenth-century health guides for medical discussions about alimentary choices. It builds on work already done by Anita Guerrini, Steven Shapin, Lucia Dacome, as well as Barbara Orland and Emma Spary, who have demonstrated that mechanical physiology did lead to changes in dietary habits among readers (and indeed, the authors) of these health guides.[9] In this chapter, I will investigate changes in advice relating to dietary choices and demonstrate that these were caused by a number of factors including the shift from Galenic to mechanical models of the body.

The health guide underwent some changes in the eighteenth century. The traditional diet, or 'regime', was not a treatise exclusively dedicated to food and drink but contained rules concerning all six of the non-naturals (air, food and drink, exercise and rest, sleep and waking, excretion and

[8] Jeremiah Wainewright, *A mechanical account of the non-naturals: being a brief explication of the changes made in humane bodies, by air, diet &c* (London: Printed for Ralph Smith, 1707), Eighteenth Century Collections Online, pp. 159–160. The national dimension to the health advice of this period has been discussed by other historians, for example Anita Guerrini, 'Health, national character and the English diet in 1700', *Studies in History and Philosophy of Biological and Biomedical Sciences*, 43, (2012), and Steven Shapin, "You are what you eat': historical changes in ideas about food and identity', *Historical Research*, Vol. 87, no. 237 (August 2014).

[9] Anita Guerrini, *Obesity and depression in the Enlightenment: The life and times of George Cheyne* (Norman, University of Oklahoma Press, 2000); Shapin, "Trusting George Cheyne"; Dacome, "Living with the chair"; Orland and Spary, *Studies in History and Philosophy of Biological and Biomedical Sciences.*

retention, and passions of the mind). Though the term diet was used to describe the habitually taken food and drink, the word 'diet' was also used to mean 'way of living', which was the original meaning of the Greek ancestor of the term, *diaita*. In fact, *diaita* was commonly used to refer to a way of living prescribed by a physician.[10] 'Diet is designed by very learned Scholars, an exact order in Labour, Meat, Drink, Sleep, and Venery', English physician Thomas Muffett (1553–1601) explained, but he was one of the earlier physicians to apply the term to 'an orderly and due course observed in the use of bodily nourishments' specifically.[11] Muffett was not alone herein. Over the course of the seventeenth century the word diet increasingly came to refer to food and drink, while the term 'regimen' was used to refer to a course or rules concerning all or most of the six non-naturals, and the term 'hygiene' fell out of use. The Latin term *regimen* derives from the verb *regere*, which means 'to lead straight' or 'to rule', and was used in English health literature to refer to a system of rules or guidelines designed to pursue a healthy lifestyle.[12] These guidelines almost always included rules relating to food and drink, but these featured alongside rules about exercise, air, sleep, retentions and excretions, and the passions of the mind.

From the mid-seventeenth century onwards, however, texts started to focus on only some non-naturals (air and aliment being particularly popular) but not all, and treatises which discussed only a single non-natural also started to appear in this period.[13] This was also true for health manuals

[10] *Merriam-Webster.com Dictionary*, s.v. "diet," accessed September 5, 2022, https://www.merriam-webster.com/dictionary/diet.

[11] Thomas Muffett, *Healths Improvement, or, Rules Comprizing and discovering the Nature, Method, and Manner of Preparing all sort of Food used in this Nation ... Corrected and Enlarged by Christopher Bennet* (London: Printed by Thomas Newcomb, 1655), The British Library, p. 1.

[12] Merriam-Webster.com *Dictionary*, s.v. "regimen," accessed September 5, 2022, https://www.merriam-webster.com/dictionary/regimen.

[13] For example, T. Byfield wrote about air, food and drink, and transpiration in his *Two discourses one of consumptions, with their cure by a new method, the other contains some rules of health* (London: Printed by Dorman Newman, 1685), Early English Books Online, while *The British Physician* treated of *our Diet, and Common Nourishment, and* of *Air* (London: Printed for J. Morphew, 1716), Eighteenth Century Collections Online; William Forster's treatise, *A treatise on the various kinds and qualities of foods: with aphorisms of health, or, rules to preserve the body to a good age* (Newcastle upon Tyne: Printed by John White, 1738) Eighteenth Century Collections Online, was about food alone, and John Arbuthnot similarly write treatises focussing on aliment in particular, rather than all non-naturals. John Arbuthnot, *An essay concerning the nature of aliments, and the choice of them, according to the different*

aimed at treating particular diseases, such as the gout or corpulency, or nervous diseases.[14] While, on the one hand, this indicates that health advice became increasingly specialised, the general 'health guide' which offered broad advice on a plethora of health-related topics and was applicable to the healthy as well as the sick, remained popular. The basic foundations of humoural physiology persisted more or less intact throughout the early modern period and into the eighteenth century, even if the terms to describe certain bodily processes changed, as David Gentilcore pointed out.[15] Ken Albala argued that many physicians simply incorporated the latest medical jargon in order to appear up to date, though he adds that many physicians did make serious attempts at harmonising the older with the newer theories.[16] As such, much dietary advice continued to recommend that readers rely on common sense in regard to choosing their aliment, avoiding ingredients which made them feel unwell but making free use of ingredients with which they were familiar and which they digested with ease. Indeed, some of these treatises advised healthy readers not to follow rules at all but to use their appetites as a guide instead.[17] Most treatises, however, were keen to get the message across that health was dependent on the careful management of diet, the success of which depended on a solid grasp of medical theory regarding both food and the body.

These developments include the appearance of health guides which offered courses of diet targeted at large groups of people, as well as an increasing tendency among certain physicians to suggest that certain foodstuffs were a generally better dietary choice than others. These treatises, and the physicians who wrote them, departed from the traditional

constitutions of human bodies (Dublin: Printed for S. Powell, 1731) Eighteenth Century Collections Online, and *Practical rules of diet in the various constitutions and diseases of human bodies* (London: Printed for J. Tonson, 1732), Eighteenth Century Collections Online.

[14] Such as: Anon., *Sure Methods of improving health and prolonging life … and maxims of health, for the bilious and nervous, the consumptive, men of letters, and people of fashion, by a physician.* 2nd edition (London: Published by Simkin and Marshall, Stationer's Court, 1827) www.archive.org; Bennet, *An essay on the gout*; Thomas Byfield, *Two discourses one of consumptions, with their cure by a new method, the other contains some rules of health* (London: Printed by Dorman Newman, 1685) Early English Books Online; and George Cheyne, *The English malady: or, a treatise of nervous diseases of all kinds* (London: Printed for G. Strahan, 1733), Eighteenth Century Collections Online.

[15] Gentilcore, *Food and Health in Early Modern Europe*, 35–36.

[16] Ken Albala, *Food in Early Modern Europe* (Westport, Greenwood Publishing Group, 2003), 227.

[17] Arbuthnot, *An Essay Concerning the Nature of Aliments*, 210–211.

approach taken by their predecessors, and this chapter will explain why this departure was possible and also why the impact of these departures on eighteenth-century dietary literature remained somewhat limited. As such, this chapter contributes to the existing research which looks into the major shifts between eighteenth-century and modern approaches to diet, particularly in trying to answer the question of when, how, and why physicians shifted from bespoke to more generally applicable dietary advice. That being said, as these developments were the result of a unique set of circumstances in the medical culture of the early eighteenth century, we need to be careful to not assume that these were precursors to later developments but rather appreciate the importance of these events in their own context.

Small and Large Particles: Feeding the Machine Body

The early eighteenth century was a period in which many different theories and models concerning the human body co-existed. New schools of thought had challenged many of the more traditional Galenic models of the body and theories about how and why this body needed to be nourished, but classical explanations remained popular throughout the eighteenth century, surviving into the nineteenth century and beyond. The iatromechanical school of thought, popular especially in the late seventeenth and early eighteenth centuries, explained the role of food in maintaining the health of the body quite differently from the traditional explanations. This is especially evident in the work of two physicians in particular, namely Bath-based physician George Cheyne (1671–1743) and Italian physician Antonio Cocchi (1695–1758), whose *Del vitto pitagorico per uso delle medicina* (1743) was translated into English by bookseller Robert Dodsley in 1745 and received with great enthusiasm.[18] Where classical theory generally described the role of food as a primary method of either maintaining or correcting the humoural equilibrium in the body, mechanical physicians such as Cheyne and Cocchi argued that the

[18] It was published as: Antonio Cocchi, *The Pythagorean diet, of vegetables only, conductive to the preservation of health, and the cure of diseases* (London: Printed for R. Dodsley, 1745), Eighteenth Century Collections Online. Ken Albala, "Insensible perspiration and oily vegetable humor: an eighteenth century controversy over vegetarianism", *Gastronomica*, 2, 3, (2002), 29.

intention of food was to replenish what the body lost during the day, as swiftly and precisely as possible, while avoiding any blockages in the tubular vessels of the body. These two different ideas co-existed, but iatromechanical explanations of the body and the role of food in maintaining the health of the body were prevalent especially in the late seventeenth and early eighteenth centuries when the writings of Cheyne and Cocchi started to appear. Both Cheyne and Cocchi embraced the idea that a healthy diet was one which was constructed to replenish, rather than (re-)balance, and their approaches to diet can be partially attributed to this rethinking of the nature and role of food.

Classical medical literature argued that all bodies consisted of a mixture of humours in which generally one of the four humours (blood, phlegm, yellow bile, black bile) dominated. Having a predominant humour was not necessarily indicative of ill health but imbalances or changes in the humoural makeup of the body could cause illness, it was believed, in which case re-balancing through the management of the six non-naturals (air, food and drink, exercise and rest, excretion and retention, sleep and waking, and the passions of the mind) was necessary. All the six non-naturals played a part in maintaining this balance, although diet was generally discussed most extensively of all in this context. Selecting the right kind of ingredients required intricate knowledge of one's own constitution, as what was wholesome for one, could be detrimental to another. For example, Cambridge physician William Vaughan (1648–1712) argued that 'Young mutton boyled and eated with opening and cordiall hearbs, is the most nourishming meat of all, and hurteth none', except for 'flegmatick persons, and those which are troubled with the dropsy'.[19] Similarly, young beef was 'naturall meat for men of strong constitutions', while the meat of young pigs was hurtful to 'old men, and idle persons', and goat was 'wholesome for sicke and weake folks'.[20] Traditionally, the right choice of meat was dependent on a range of individual factors, including constitution, particular illnesses, age, activity levels, and one's general 'strength' or health.

In this humoural system, the relative wholesomeness of food depended not only on the body of the consumer but also on the food itself, because

[19] William Vaughan, *Approved directions for health, both naturall and artificiall derived from the best physitians as well moderne as auncient.* 2nd edition (London: Printed by T. S[nodham] for Roger Lackson, 1612), Early English Books Online, 36.

[20] Ibid., 36–37.

foodstuffs were believed to possess humoural qualities as well.[21] Foodstuffs could be hot, cold, moist, and dry, or temperate, and all in different degrees, making them proper for people with one constitution, but improper for another. As such, to either preserve one's current humoural balance, or alter it, people had to consider the humoural qualities of everything they consumed. Almost all physicians in this period emphasised the importance of adjusting diet to one's personal constitution. Muffett explained: 'As there is a certain temperature and distemperature of our bodies, so likewise there is in meats; that temperate bodies should feed of their likes; and distempered of their contraries'.[22] For example, a sanguineous constitutions, which abounded in hot and wet humours, required food which was of a cold and dry nature to correct the overly hot and wet state of the body—this is generally referred to as a doctrine of opposites. Early modern physicians provided long lists of all popular food items and informed readers for which constitution individual ingredients were appropriate or not appropriate. Since the healthiness of these food items was relative to the constitution of the body which consumed them, physicians rarely deemed any single ingredient or food group to be 'healthy' or 'unhealthy'. [23] Personal taste also had to be accounted for: 'for some have loathed and abhorred Cheese, others Wine, others Garlick &c', and physicians often expressed the belief that fondness or dislike for certain ingredients was the body's own guide for selecting which ingredients were suitable or unsuitable for it.[24] For these reasons, many early modern physicians rejected the very notion of general dietary rules.

Traditionally, dietary advice did not generally promote one foodstuff to vilify another, though some treatises debated the benefits or drawbacks of new foodstuffs such as tea and coffee, while others discouraged the use of tobacco or alcohol. Instead, many traditional health guides provided guidance concerning the right quantity and timing of the meals, along with long lists of ingredients, the qualities of these ingredients, their effects on the body, and to whom these ingredients might be wholesome or unwholesome. Health guides were not generally intended to function as a vehicle for laying down rules for all to follow: they were a guide to aid readers in

[21] Ibid., 35.

[22] Muffett, *Healths Improvement*, 32.

[23] Castor Durante, *A Treasure of health by Castor Durante Da Gualdo, physician and citizen of Rome, wherein is shewn how to preserve health and prolong life* (London: Printed for William Crook, 1686), Early English Books Online, 51.

[24] Durante, *A Treasure of Health*, 52.

constructing their own rules by providing them with the necessary information about the qualities of food. Sometime over the course of the eighteenth century, however, this started to change. While seventeenth-century physicians concerned themselves with the unique dietary needs and preferences of individuals, some early eighteenth-century physicians started to publish dietary rules they considered to be applicable to larger groups of people. To some extent, these physicians made fewer concessions to personal appetites and prescribed aliments of set quantities and qualities to their whole readership. Such a change is considerable, and research into the origins, causes, and development of this shift may help us better understand attitudes to food and health in the past.

The iatromechanical school introduced some new terms and theories to the early eighteenth-century health guide. Terms such as 'fibres', 'tension', and 'elasticity' were now some of the many factors physicians had to consider in the diets they prescribed. Interest in the solid parts of the body started to develop in the seventeenth century and seemed to have reached their peak in the early to mid-eighteenth century, as mechanical conceptions of the body were popularised by the theories and works of Rene Descartes (1596–1650), Isaac Newton (1643–1727), Herman Boerhaave (1668–1738), and Julien Offray de La Mettrie (1709–1751). According to these mechanical theories, the solid fibres of the body played an active role in the distribution of fluids through the body. Iatromechanical theory imagined the body to function as a hydraulic machine consisting of many pipes and strainers through which fluids containing nourishment was filtered into increasingly small particles, which would then either be assimilated into the body or be cast out as waste matter through the excretory system. Bath physician George Cheyne explained his understanding of the machine body vividly in his *Essay on Regimen* (1740), where he described the body as:

> Nothing more but a Compages or Contextures of Pipes, an hydraulic Machin, filled with a Liquor of such a Nature as was transfus'd into it by its Parents or is changed into by the Nature of the Food it is nourish'd with, and is ever afterwards good, or bad, or indifferent, as these two Sources have sent it forth. The Primæ Væ, or the alimentary Tube, is a Common Sewer, may be foul'd or clean'd in various Manners, and with great Facility; it is wide, open and reasonably strong, in most Animals: But from the Intestines into the Habit, and into the sanguiserous Tubes, into the Glands and Bowels, the Passages from thence are so streight and narrow, that they are

> but infinitesimal Pores, rather than Orifices, Strainers and Searches, rather than Mouths of Tubes, and are scarce ever visible or perceptible in a healthy State: and the Supply and Nutriture of the Fluids and Solids must pass under the Form of a Steam or Vapour through them, and not the Form of a Liquor.[25]

Food was first broken down by the widest pipe, the gastrointestinal tract (the oesophagus, stomach, and intestines), but particles would travel through increasingly narrowing passages, which would require nourishment to be broken down into smaller and smaller particles, until what was left was little more than 'Steam' or 'Vapour' which would eventually be cast out of the body through the pores of the skin or through the mouth. The primary way to keep such a hydraulic machine body healthy was through ensuring that no blockages disrupted the flow of the fluids and the steam, and through ensuring that the 'Juices' which flowed through these pipes, contained the nourishment necessary to keep the solid parts of the machine, the tubular vessels, in a state of health.[26] Similarly to Cheyne, Italian physician Antonio Cocchi described the healthy functioning of the body as 'the perpetual and equable Motion of a large Mass of Liquid distributed through innumerable continued Channels, which dividing into Trunks and Branches, are reduc'd in their Extremities to an inexpressible Degree of Minuteness'.[27]

These pipes, often also referred to as 'vessels' or 'channels', consisted of minuscule interwoven fibres, and they contained and propelled forth all the fluids of the body but could easily become blocked or damaged by alimentary particles. Consequently, the correct maintenance of these fibres of the body gained increased importance in early eighteenth-century discussions of the non-naturals, including diet.[28] The fibres of the body were understood to be under a constant state of tension, or elasticity, in order to push or 'propel' forth the fluids contained within them. This tension needed to be maintained in order to maintain due circulation of the body's

[25] George Cheyne, 'A Philosophical and Practical Essay on the General Method and Medicins' *An Essay on Regimen* (London: Printed for Dan. Browne, 1753), ii-iii.

[26] Ibid.

[27] Antonio Cocchi, *The Pythagorean diet, of vegetables only, conductive to the preservation of health, and the cure of diseases* (London: Printed for R. Dodsley, 1745), Eighteenth Century Collections Online, 41.

[28] This focus on the fibres of the body also had a tremendous effect on exercise advice, as we will discuss in Chaps. 3 and 4.

fluids. It was argued that with every single action of the body, the fibres of the body weakened over time until they ultimately lost tension and became unfit to perform their functions.[29] The only way to slow or undo the damage caused by the very functions of the body was by restoring the tension of these fibres. The health of this machine body was therefore preserved through ensuring two things: first, the constant and uninterrupted flow of the fluids through the vessels, and second, the maintenance of the tensions or elasticity of the vessels which propelled forth these fluids.

This hydraulic concept of the body was accompanied by a new way of imagining food and the role that aliment played in the maintenance of the machine body. To keep the fluids of the machine body in a constant, uninterrupted flow, physicians of the late seventeenth and early eighteenth centuries discerned between foods which either promoted or hindered the circulation of the fluids, promoting consumption of the former, and advising against the latter. Foods which promoted the circulation of the fluids were those which were easily digested and unlikely to cause blockages in the vessels. The 'digestibility' of aliment has long been a feature of dietary discussions but became a focal point in this period especially, and many physicians attempted to classify aliment according to their perceived ease or 'lightness' of digestion.[30] London physician and prolific medical author Everard Maynwaring (c.1629–1713) described heavy meats as 'such as are more dry, hard, solid, sense, gross, course and tough', requiring a longer time to digest, before they were passed on from the stomach into the rest of the body.[31] Light aliment, on the other hand, was 'soft and tender' and sooner yielding to digestion.[32] For many physicians, whether heavy or light food was the more healthful option continued to depend on the

[29] Wainwright, *Mechanical account*, 149; Forster *A treatise on the various kinds and qualities of foods*, 3–4.

[30] See, for example, Everard Maynwaring, *The method and means of enjoying health, vigour and long life adapting peculiar courses for different constitutions, ages, abilities, valetudinary states, individual proprieties, habituated customs, and passions of mind* (London: Printed by J.M. for Dorman Newman, 1683) Early English Books Online, 54–55.

[31] Maynwaring, *The methods and means* 1683, 54. Note: 'meats' here means all solid aliment. For a comprehensive overview of Maynwaring's career, medical writings, and his role in late seventeenth-century medical debates surrounding experimental medicine, see Jonathan Barry, 'The 'Compleat Physician' and Experimentation in Medicines: Everard Maynwaring (c.1629–1713) and the Restoration Debate on Medical Practice in London', *Medical History*, 62, 2 (April 2018), 155–176.

[32] Ibid., 55.

body which consumed it.[33] For example, in *The Buckler of Bodilie Health* (1630), it was advised that while 'Grosse and viscuous meate causeth obstruction', those who had a strong constitution could consume them 'boldly without hurte'.[34] 'Light meate' was unsuitable to 'leane people, and of a hote complexion' but 'fitting for growne and grosse bodyes' and 'for flegmaticks, and for these who are of a weake stomach'.[35]

Other physicians, however, especially in the early eighteenth century, theorised that food which was digested with relative ease was the right choice for a great many people and recommended a diet consisting of such 'light' and 'digestible' foods to readers of their health guides. George Cheyne ascribed easy digestibility especially to vegetable food, while most meat was described by him as less digestible and therefore less suitable for his readers. According to Anita Guerrini, Cheyne was a 'household name' by the 1720s, following the publication of his *Essay on Gout* (1720) and *Essay of Health and Long Life* (1724).[36] Though educated in Edinburgh, Cheyne travelled to Bath for his health, that being the site of the by then increasingly famous health spa, where he consequently set up his practice. A major influence on Cheyne's work was his mentor, physician Archibald Pitcairne (1652–1713) of Edinburgh. Pitcairne was well-versed in mathematics as well as medicine and had argued that only mathematics could provide a basis of certainty for medicine.[37] This preference for mathematical approaches to medicine can be discerned in Cheyne's work as well, which advised that a healthy diet could be determined through measurement and calculation. Some foods broke down into smaller parts than others, and these smaller parts fitted through the body's finer vessels with greater ease than the larger particles. Finding out which foods broke down smaller was a matter of experiment, and deciding on a healthful diet a matter of establishing rules based on those experiments.

Cheyne's advice skewed towards promoting the 'light' easily digestible foodstuffs over the 'heavier' aliment. In fact, one of the primary sources of chronic distempers, he told his readers, was the 'over-largeness' of the

[33] Maynwaring, *The methods and means*, 56.

[34] John Makluire, *The Buckler of bodilie health whereby health may bee defended and sicknesse repelled*. 2nd edition (Edinburgh: Printed by John Wreittoun, 1630), Early English Books Online, 37.

[35] Ibid.

[36] Anita Guerrini, *Obesity and Depression in the Enlightenment: The Life and Times of George Cheyne*. Oklahoma: Oklahoma University Press, 2000, xix.

[37] Ibid., 28–30.

particles of the food they consumed.[38] The cure for all of these distempers was therefore to follow a diet consisting of aliment which was easy to digest, and Cheyne's *Essay of Health and Long life* provided some short-cuts to determining which foods met this criterion. 'Vegetables and animals close to maturity are the lightest of digestion', Cheyne advised. 'Spring vegetables such as asparagus, strawberries, salad, is easier digested than Pears, Apples, Peaches'; and among animals, Cheyne found that meat from animals such as poultry, hares, and sheep were more tender and easier digested than cows and horses, because 'their Parts cohere less firmly' and were therefore more easily broken down by the body into particles small enough to pass through the body's narrowing vessels without causing blockages.[39] Cheyne reasoned it was better to err on the side of caution by choosing aliment which was easily digested and avoiding that which was digested with difficulty.[40] 'The Meat of England is generally animal Substances', Cheyne observed, which consisted in 'closely compacted Parts' which were hard to digest.[41] Not all meat was the same: the fitness or unfitness of animal food depended on the size, age, and rearing of the animal.[42] Not all vegetable food was alike either: their cohesion and chemical compounds depended on their kind, size, and the season in which they were consumed. Not all flesh was difficult to digest, just as not all vegetable food was easily digested. Nevertheless, Cheyne advised his readers to generally avoid the former and consume more of the latter. First, he asserted that 'the proper Food appointed for Animals was easier to digest than the animals themselves', and animals which fed on vegetables were easier to digest than animals which lived on animals, creating a food pyramid of digestibility at which vegetable food came out on top, flesh of vegetable-consuming animals in the middle, and flesh of animals which had lived off flesh at the bottom.[43]

In doing so, Cheyne was generally more willing than other physicians to promote certain foodstuffs over others. This is especially the case in Cheyne's praise of the qualities of milk. Aside from its general emphasis on

[38] Cheyne, *An Essay of health and long life*, 19.

[39] Ibid., 22.

[40] George Cheyne, *An Essay of health and long life*. London: Printed for George Strahan, 1724. Eighteenth Century Collections Online, 27.

[41] Ibid., 20. In early modern usage, the term 'meat' did not exclusively refer to animal substances but to all solid aliment.

[42] Ibid.

[43] Ibid., 24.

vegetable foods over meat, Cheyne's diet was perhaps most famous for its promotion of milk as a key ingredient to his 'light' or 'low' regimen. In fact, Cheyne's was often referred to, by himself as well as others, as the 'milk diet'. Cheyne described milk as 'white Blood already made, adjusted to all the Meander of Circulation', meaning that it needed very little work from the body to turn milk into nourishment for the body.[44] Cheyne argued that the natural purpose of milk was 'for the nourishing and fortifying of weak and tender animals', which also made it especially suitable for tender and sick persons.[45] Cheyne's definition of tender persons was quite broad however, and he recommended the consumption of copious amounts of milk to many of his readers. According to Cheyne, milk was ready-made chyle, a happy medium between meat and vegetable in that it was nourishing but of the easiest possible digestion as it was already prepared 'by the hand of Nature'.[46] Cheyne argued that even persons who had a natural aversion to milk ought to find ways to consume it regardless. He recommended people dilute the milk and mix it with other foodstuffs to reduce the curdling that took place in the stomach. Avoiding milk, he argued, was 'an infinite loss to them', as it was an efficient antidote to all chronic distempers.[47] While it was not unusual for physicians to discuss what they perceived to be the benefits of any particular foodstuff, Cheyne went further than most in suggesting that milk was a universally healthy ingredient.

Similarly to Cheyne, Italian physician Antonio Cocchi located the benefit of vegetable food in the easy of their digestion. In his essay on the benefits of Pythagorean diet, Cocchi argued that the parts of plants had a much more tender, less cohesive texture than meat, which meant that vegetable food was more easily broken down in the digestive tract than meat.[48] Furthermore, Cocchi argued that plant-based aliment 'abounded' in water and was moderate in oily fluid, the mixture of which resulted in an 'incorporating and dissolving Juice', subtle enough to facilitate the fluid distribution into the numerable vessels, while particles which were 'hard and weighty' would lead to blockages in the channels which needed

[44] George Cheyne, *The natural method of cureing the diseases of the body, and the disorders of the mind depending on the body*. London: Printed for George Strahan, 1742. Eighteenth Century Collections Online, 127.

[45] Ibid.

[46] Ibid., 126.

[47] Ibid., 130.

[48] Cocchi, *The Pythagorean diet*, 48.

to be kept open.[49] Due to their being easily broken down into smaller parts, vegetable food was argued to be more quickly distributed through the minuscule vessels of the body, to ultimately be cast off as perspiration through the mouth or the pores of the skin.[50] Food items were thus divided into wholesome or unwholesome along these lines of digestibility, rather than whether they were animal or vegetable per se. For example, Cocchi advised that spices were hot and dry and best avoided, while fresh herbs were beneficial. Any food which was dried, whether vegetable or not, was equally vilified for the same reasons as animal flesh.[51] Just like Cheyne, Cocchi believed that milk was uniquely positioned between vegetable and animal food, still containing the vegetable qualities of the plants the animal fed on while having been 'triturated' by the animal into a 'most light, good and simple Aliment', which was readily converted by the substance of the human body.[52] Cheese, for the same reasons as milk, was also good, as long as it was 'fresh and new': cheese which was hard and dry should be used only very occasionally and then only in combination with aliment which offset its bad qualities.[53] Neither Cocchi's Pythagorean nor Cheyne's milk diet was therefore strictly about vegetarianism as much as they were about promoting those foods they had found to be most easily digested and best promoted the easy circulation of nourishment through the body that was imagined to function much like a hydraulic machine.

It may be due to this focus on the shared qualities of food as being 'digestible' that physicians Cheyne and Cocchi were able to recommend their diets to groups of people, rather than just individuals. Their approach quickly drew criticism from others, however. One particular periodical published around the same time as Cheyne's *Natural Methods* criticised physicians, who 'in their Mathematical Calculations only have regard to proportionable Difference, or Largeness and Smallness of Parts', and were mistaken in their assumptions that such a mathematical approach to regulating diet was practical. All aliments, the periodical argued, whether meat or vegetables differed in their 'Virtues, Quality, and Strength', according to their different 'Natures, Climes, Ages and Nourishment' and so a course of diet could never be reduced to some general rules about the

[49] Ibid., 50.
[50] Ibid.
[51] Ibid., 52.
[52] Ibid., 53.
[53] Ibid., 56.

cohesion of alimentary matter, these critics argued.[54] These treatises tended to emphasise the importance of individual factors such as constitution that Cheyne and Cocchi seemed to be moving away from. Cheyne's and Cocchi's approach was therefore far from the norm for physicians of this period.

Nourishing and Perspirable: The New Food Qualities

Physicians, including Cheyne and Cocchi, also started to ascribe to aliment qualities related to the degree to which they nourished the body. What proportion of food became absorbed by the body for nourishment or excreted as waste matter became a topic of great interest in the early eighteenth century. To support their iatromechanical theories, Cocchi and Cheyne looked back at the experiments undertaken by Santorio Santori in the late sixteenth century, who, following a 30-year weighing trial, had attempted to discern which aliments were harder or easier of digestion and more or less conductive to perspiration. Santorio was born in Capodistria (today Koper in Slovenia) on 29 March 1561 and graduated from the University of Padua in philosophy and medicine in 1585.[55] He developed his interests in quantification and experimental medicine early on in his career, and Fabrizio Bigotti has estimated that he may have begun experimenting on himself and others as early as 1584.[56] As Bigotti recently argued in a comprehensive volume on Santorio, the *Medicina statica* (1614), in which Santorio published his findings, became a textbook for generations of physicians and was republished several times between the late seventeenth and early eighteenth centuries.[57] The first English editions of *Medicina statica* appeared in 1676.[58] It was later also translated by medical writer and student of Newtonian philosophy John Quincy (unknown to 1722), as *Medicina statica, being the Aphorisms of Sanctorius,*

[54] Anon., *The British Physician Treating of our Diet, and Common Nourishment*, London Number 2, June 12th, 1716.

[55] Fabrizio Bigotti, "Introduction", in *Santorio Santori and the Emergence of Quantified Medicine, 1614–1790,* eds. Jonathan Barry and Fabrizio Bigotti (London: Palgrave Macmillan, 2022), 5 and 8.

[56] Ibid., 9.

[57] Ibid., 2.

[58] Santorio Santori, *Medicina Statica, or, Rules of health in eight sections of aphorisms originally written by Sanctorius; English'd by J.D.* (London: Printed for John Starkey, 1676), Early English Books Online.

translated into English, with large Explanations (1712), which would be revised and republished another five times until 1738.[59]

Similarly to Santorio, eighteenth-century iatromechanical physicians set about trying to determine how to achieve a perfect balance between *ingesta* (the sum of food and drink taken in) and *excreta* (the sum of all the body's excretions). Increasingly, it was argued that perfect health and long life were achieved not through balancing the humoural equilibrium but by ensuring that one's dietary intake answered perfectly for the matter daily excreted by the body and the daily repair of the solids. Cheyne was especially explicit in arguing that this was the very purpose of eating. 'Food was design'd', he argued, to supply 'the Expenses of Action and Living'.[60] As long as the supply of food was equal to the expense of living, he argued, people would remain free from chronic illnesses.[61] Not all that was eaten was assimilated into the body and used for that repair, however. A proportion of all aliment was termed excrementitious (not suited for assimilation) or superfluous (excess matter) and needed to be cast off through the excretory system. How much the body retained or excreted of the aliment it consumed was unknowable until Santorio invented his weighing chair, but equipped with this method of measuring, early eighteenth-century physicians set about establishing exactly what was and what was not absorbed by the body.

A period of experimenting commenced in Britain as well as other parts of the world. In 1720, the results of experiments undertaken by Scottish physician James Keill were published by John Quincy as *Medicina statica Brittanica*, in a volume which also contained Santorio's work.[62] Keill argued that according to his measurements, the human body lost about four pounds of weight every day through its excretions. The proportion of daily aliment needed for that body, he concluded, was therefore also four

[59] Santorio Santori, *Medicina Statica: being the aphorisms of Sanctorius, translated into English with large explanations. Wherein is given a mechanical account of the animal economy, and of the Efficacy of the Non-Naturals, ... by John Quincy* (London: Printed for William Newton, 1712), Eighteenth Century Collections Online.

[60] Cheyne, *natural methods*, 8. Similar arguments were made by Forster, *A Treatise on the Various Kinds and Qualities of Food*, 3–4, and Wainwright, *A Mechanical Account*, 149.

[61] Cheyne, *An Essay of health and long life*, 19.

[62] Santorio Santori, *Medicina Statica: being the aphorisms of Sanctorius ... To which is added Dr. Keill's Medicina Statica Britannica, with comparative remarks ... by John Quincy*. 2nd edition (London: Printed for W. and J. Newton, 1720), Eighteenth Century Collections Online.

pounds, to make up for the loss and keep the body in a constant state of equilibrium between the *ingesta* and *excreta*.[63] As a result, some dietary rules became matters of mathematical equations based on quantitative measurements.[64] By introducing the aphorisms of Santorio in the vernacular, static medicine entered the wider early eighteenth-century English dietary culture, the impact of which is apparent in not only the numerous medical re-trials of the Santorian experiment which took place in the early eighteenth century but also by some physicians' new recommendations to make dietary decisions based on measurements of body weight.[65] Monitoring shifts in the weight of the body was a seemingly straightforward method of finding out whether one ought to increase or decrease one's daily intake of food. 'And', Santorio had argued, 'if it be known exactly how much Food is convenient every Day, such a Person may easily preserve his Health'.[66] Such an approach to diet was novel, as previously physicians had always recommended patients gauge the correct amount of food by sensations of hunger, fullness, and signs of digestive discomfort.[67]

It is worth noting here that these physicians' interest in perspirable or 'lighter' food over more nourishing aliment ought not to be conflated with modern low-calorie diets. There has been a tendency among historians to assume that these discussions involving body weight and 'heavy' or 'light' foods prefigure modern dietary practice motivated by concerns with overweight. Ken Albala, for example, suggested that medical professionals of the seventeenth and eighteenth centuries had been responsible for a 'gradual shift in meaning of the word diet from a general programme

[63] Keill, *Medicina Statica* (version 1720), 336.

[64] On a comprehensive discussion about dietary quantities in early modern health guides, see this chapter.

[65] Both Cheyne and Arbuthnot suggested the use of the 'Santorian chair' to determine not only what quantity but also what quality of aliment was suitable. John Arbuthnot, *An essay concerning the nature of aliments, and the choice of them, according to the different constitutions of human bodies* (Dublin: Printed for S. Powell, 1731), 198–199, Eighteenth Century Collections Online; George Cheyne, *An essay on Regimen*, (London, Dan. Browne, 1753), Iiii.

[66] Santorio Santori, *Medicina Statica: being the aphorisms of Sanctorius, translated into English, with large explanations. To which is added, Dr. Keill's Medicina Statica Britannica … By John Quincy.* 5th edition (London: Printed for T. Longman, 1737), 186, Eighteenth Century Collections Online.

[67] And indeed, many physicians continued to stress the accuracy of these physical sensations over cumbersome and unreliable measurements, as I discuss in this chapter.

of health to a regimen specifically designed to cure excess weight'.[68] High body weight in itself was not a concern among eighteenth-century physicians. As will be discussed later, both the execution of and motivations for eighteenth-century body weighing practices differ considerably from modern weight-watching practices in that they concerned themselves with shifts in the balance between the *ingesta* and *excreta*, and not with absolute weight or concerns with corpulency, which was considered an illness defined by excessive fatness, but not by high body weight.[69] It is therefore important that we do not assume these practices to be a precursor to today's weight-watching habits.[70]

Traditionally, food was ascribed humoural qualities such as hot or cold, dry or moist, but physicians Cheyne and Cocchi described food as 'nourishing' or 'perspirable'. Though Santorio was not the first to discuss insensible perspiration, his *Medicina statica* ascribed an importance to the matter of perspiration and the role food played therein that was not present in older health guides. Insensible perspiration was a waste product of aliment, created in the very last stage of digestion, when the nutritive matter had been rendered so small that it would be cast out through the smallest openings of the body: the pores. How much a person daily perspired depended on several factors, including climate and temperature, one's exercise levels, and constitution, but above all, it depended on the choice of aliment. Certain kinds of food create more waste product than others, physicians argued. In his treatise, Santorio had discussed certain food types which were either 'very perspirable' or 'of little perspiration'. Aliments producing little perspiration were those which were understood to be 'nourishing'. Nourishing foods replenished the body and fattened it. Examples of food which was nourishing included pork, eel, nuts, and all sort of other aliment generally described as 'fatty' (if animal) or 'oily' (if vegetable). Though these foods replenished and added bulk to the body, they were also digested with some difficulty and could cause obstructions in the body if not properly digested. Once this aliment was finally digested, however, the particles were assimilated into the fluids and solids of the body, leaving but little waste to be expelled through insensible perspiration.[71] 'Lighter' aliment such as chicken and many kinds of vegetables

[68] Albala, 'Weight Loss in the Age of Reason', 169.
[69] See Chap. 6 of this book.
[70] This argument will be fully presented in Chaps. 4 and 5 of this book.
[71] Quincy, *Medicina Statica* (version 1720), 168.

resulted in considerable perspiration. Contrary to nourishing foods, foods which were 'perspirable' were quickly and easily digested and broken down into small parts to ultimately be cast off through the pores as insensible perspiration.[72] They added but little weight to the body and consequently did not replenish the body as much as nourishing foods did.

Santorio condemned foods which did not perspire or reduced the perspirability of other foods: 'Pork and Mushrooms *are bad*, both because they do not perspire themselves, and because they hinder the Perspiration of other Things eat[en] along with them'.[73] Foods which did not perspire, he argued, occasioned 'Obstructions, Putrefactions, Lassitudes, Sadness and an Increase of Weight', while foods which were perspirable would preserve a person from 'Heaviness and Weariness'.[74] As such, it was Santorio who first implied that aliment which did not result in (much) perspiration could be considered less wholesome than aliment which did. It was not as straightforward as perspirable foods being absolutely healthy and nourishing foods absolutely unhealthy, however. Santorio admitted that perspirable foods were not as efficient at 'recruiting the decayed Strength' as were foods of much nourishment and little perspiration.[75] For people in need of much nourishment, such as labourers, or people weakened by illness, most perspirable foods would not supply sufficient to make up for their loss of strength. That being said, Quincy added that giving food of 'substantial Nourishment' to 'weak and decay'd Persons' was not always correct, because these foods also required great digestive strength to break down their cohesive particles and mix these with the blood—strength which persons weakened by illness, for example, would not possess.[76] Consequently, whether aliment of a more perspirable quality was more or less appropriate than aliment which was more nourishing was a question which even Santorio himself could only determine with reference to the body that consumed this aliment.

Many other physicians found fault with Santorio's tendency to focus on food quantities over qualities in his aphorisms. 'Tho Sanctorius gives a Rule to measure that quantity of Food', Wainwright argued, his aphorisms were 'faulty in several Respects', 'for the same quantity of some sorts of

[72] Ibid.
[73] Santorio, *Medicina Statica* (version 1737), 182. Emphasis my own.
[74] Ibid., 196.
[75] Ibid., 205.
[76] Quincy, *Medicina Statica* (version 1737), 205.

Meat and Drink, will almost expel double the quantity of Perspirable Matter, in the same time, that other sorts will do it in'.[77] In other words, the quantity of perspiration that aliment wrought was not a good indicator of whether individual foodstuffs or a course of diet was healthy or not. The early eighteenth-century experimenter James Keill (1673–1719) came to the same conclusion. He explained: 'It requires good Attention to distinguish between Things that lessen Perspiration, by being uncapable of Reduction … to a fineness suitable for such Discharge, and such as lessen Perspiration by their fitness for Lodgement in the Habit, and making Matter for Nourishment'.[78] The first, he continued, were substances 'hard and viscid' that could not be broken down finely enough to be expelled through the pores and so were excreted through urine or stool instead. The latter, however, could be broken down into small parts but, were of such an easily digestible nature, that they were also easily assimilated into the body 'and make a Part of the Passages themselves', which of course lessened the quantity of waste product which needed to be perspired.[79] Certain foods, such as oysters, were of such easy assimilation that they produced barely any waste at all, complicating any attempt at distinguishing wholesome from unwholesome aliment based on how much or how little perspiration they produced.[80]

To ascertain which foods provided more or less nourishment, or more or less perspiration, eighteenth-century physicians such as James Keill but also Bryan Robinson (1680–1754) and William Stark (1740/41–1770), set about experimenting on the effects of consuming food on their bodies.[81] These experiments tended to focus on food quantities measured in weight and how these compared to the daily *excreta* of the body, but the experimenters made careful note of what they had consumed and how this affected their bodies. Keill indicated in his notes that he ate 'Water-gruel',

[77] Wainewright, *A mechanical account of the non-naturals*, 152.

[78] Keill, *Medicina Statica* (1728), 339–340.

[79] Ibid., 339.

[80] Ibid.

[81] James Keill, *Essays on the several parts of the animal oeconomy* (London: Printed for George Strahan, 1717), 231, Eighteenth Century Collections Online; Bryan Robinson, *A dissertation on the food and discharges of human bodies* (Dublin: Printed by S. Powell, 1747), Eighteenth Century Collections Online; William Stark, *The works of the late William Stark, M.D. consisting of clinical and anatomical observations, with experiments, dietetical and statical, revised and published from his original Mss by James Carmichael Smyth* (London: Printed for J. Johnson, 1788), 89, Eighteenth Century Collections Online.

which was a dish popular among other 'temperate men' such as the Italian Luigi Cornaro (1484–1566), but he also mentioned oysters, buttered wheat and mushrooms.[82] The choice of these was discussed, especially when there had been some shift in Keill's weight. For example, when he specified that he had eaten mushrooms on 27th September, Keill's morning weight had been 169 and his evening weight 171, which to Keill could have indicated a level of retention very much in line with what Santorio had argued mushrooms to cause.[83] On the 24th and 26th September, Keill noted that he had perspired less in the night than he usually did and soon decided that this was due to his supping on oysters, which led him to argue in his aphorisms that oysters gave 'the greatest Nourishment' and that they were the 'least perspirable of all Food'.[84] Keill did not determine whether mushrooms or oysters were healthful or unhealthful based on these findings, but they would be used by other physicians to suggest certain foods over others to their readers. Quincy, for example, warned that Keill's findings indicated that oysters ought not to be indulged in by those suffering from Plethora or whose digestive faculties were weak, the latter of which was believed to apply to a large proportion of physicians' readership, as we have seen in Cheyne's treatises.[85]

In his *Natural Methods*, Cheyne recalled an experiment conducted by a colleague not long after Keill published his book:

> Mr. Geoffroy has given a Methods for determining the Proportion of Nourishment, or true Matter of the Flesh and Blood, contain'd in any sort of Food. He took a Pound of Meat, that had been freed from the Fat, Bones and Cartrilages, and Boil'd it for the determin'd Time in a Close Vessel with three Pints of Water; then pouring off the Liquor, he added the same Quantity of Water, boiling it again for the same Time ... Then putting all the Liquor together, and filtrating, to separate the too gross Particles, he evaporated it over a slow Fire, till it was brought to an Extract of a pretty

[82] Keill, *Essays on the several parts of the animal oeconomy.*

[83] Santorio, *Medicina Statica* (version 1676), 78–79. According to Santorio, mushrooms 'do not perspire' and prevented other foodstuffs from perspiring if they were eaten together with mushrooms. Upon the eating of mushrooms, the body will perspire less than usual 'by a third part of a pound', Santorio argued.

[84] Keill, *Medicina Statica* (version 1720), 339; Keill, *Essays on the several parts of the animal oeconomy*, 231.

[85] Keill, *Medicina Statica* (version 1720), 339; Keill, *Essays on the several parts of the animal oeconomy*, 240.

> moderat Consistence. This Experiment was made upon several sort of Food, the Result of which may be seen in the following table.[86]

According to this table, 'white, young, tender animal Food' (such as chicken, pheasant, partridge, and mutton), as well as bread, milk, and vegetables, were 'the best and most effectual' substances for 'Nutrition, Accretion' and maintaining the smooth circulation of the fluids, just as Cheyne had argued in his treatises.[87] Whether Cheyne had based his arguments on experiments such as these is not entirely clear, but it does show that experiments which measured nourishment or perspiration from aliment were used to support arguments made in dietary advice, especially relating to choice of aliment. The quantification of dietary advice in this period was therefore not the only effect that these experiments had on health treatises. Discussions about the qualities of aliment also underwent significant changes. Increasingly, physicians started to ascribe to food qualities such as 'digestible', 'perspirable', 'heavy', or 'light'. Physicians started to recommend courses of diet which particularly involved such aliment that promoted perspiration and nourished but little, as we have seen in the treatises by Cheyne and Cocchi mentioned before. They did not claim that these were necessarily healthier in all cases, but the division of aliment into perspirable or nourishing, 'heavy or 'light' foods does seem to have allowed for a level of generality that had not been possible under the Galenic system in which food could be hot or cold, moist or dry, or temperate, to different degrees.

For the more chemically inclined physicians, such an approach to diet was unthinkable. People could not be divided into groups requiring either nourishing foods or perspirable foods. Utrecht-based physician Edward Strother (unknown to 1737) explained the blood as 'a medley of Salt, and Sour, and a thousand other Materials', and in every body these materials occurred in different proportions, the characterisation of which was still referred to as a constitution.[88] The constitutions only diversified since the writings of Galen, as eighteenth-century physicians described constitutions which were oily or saline, which could be further subdivided, since, for example, a saline constitution could be acid, alkaline, or muriatic, 'according to the Difference of Salts which occasion it'. It was therefore 'a

[86] Cheyne, *The natural method*, 55.
[87] Ibid.,56.
[88] Strother, *An essay on sickness and health*, 30–31.

standing Rule', Strother argued, that 'our Meats and Drinks ought to consist, as nearly as can be, of Particles contrary to the Cause of the Disease reigning, or contrary to the Temperament, if they be valetudinary and sickly'.[89] Scottish physician John Arbuthnot (1667–1735) explicitly argued that the whole discussion about choosing between animal and vegetable food was a moot point, because 'in the Enumeration of Constitutions ... there is not one that can be limited and restricted by such a Distinction'. Furthermore, he argued, 'there is not a general alimentary Quality in which all Vegetables agree; there are Vegetables, acid, alkaline, cooling, hot, relaxing, astringent, acrid, and mild'.[90] In other words, chemical theory of aliment and the body did not allow for the same level of standardisation as mechanical theory did.

A strict distinction between mechanically and chemically inclined physicians is of course impossible: theories of both schools of thought were generally used in tandem for physicians to demonstrate their knowledge of all the latest theories and practices. Cheyne did use chemical theories to explain matters of digestion and to a degree considered the chemical compounds of aliment to determine whether it was wholesome or not. The difference between physicians such as Cheyne and Strother, for example, is a matter of emphasis and degree, with the former selecting aliment for the purpose of solving mechanical issues with the body (maintaining the motion of the circulation by preventing blockages) and the latter, chemical (by balancing the proportion of matter to maintain a chemical state of the blood). While Cheyne did not fail to consider the salt or acid contents of meat and vegetables, ultimately these chemical analyses were used to justify a mechanical approach to curing and preserving the body.

These new terms and food qualities ultimately did not overhaul the whole of the eighteenth-century dietetics nor commence a progression towards modern-day diet culture. Most physicians continued to advise that there was no generally wholesome or unwholesome diet and that the qualities of individual food items needed to be assessed in relation to the qualities of the body consuming them. Though theories about the human body and the role that food played in maintaining the health of that body were changing, physicians were constructing and disseminating dietary advice within a medical culture which was traditional in many aspects. It is therefore important that we do not overstate the shifts that we can see in

[89] Ibid.

[90] Arbuthnot, *An essay concerning the nature of aliments*, 211–213.

dietary advice of this period. The shift of discussion to the idea of 'replenishing' from 'correcting' or 'balancing' was a matter of degree, and even Cheyne did attribute some importance to personal constitutions and circumstances in determining the correct diet for his patients. That said, we would be doing Cheyne and Cocchi a disservice by hastening to conclude that not all that much changed in this period. The level of criticism these physicians received is indicative of the significance of Cheyne and Cocci's approach to diet, both in their focus on dietary quantities and in their attempts to categorise aliment according to 'new' qualities relating to nourishment and perspiration.

Diet for a Target Audience

An important aspect of the dietary advice offered by Cheyne and Cocchi is that both physicians focussed on curing not an individual but a distemper they considered to be common to many people. Eighteenth-century medical texts increasingly specialised in particular illnesses and included dietary advice designed to cure these particular illnesses. This has already been noted by Michael Stolberg, who demonstrated that contrary to what is still often assumed, early modern medicine was not solely or even primarily based on a physiological understanding of disease as a state of imbalance between the natural humours and their qualities. This may have been true for the strictest Galenic physicians, Stolberg points out, but medical practice was already in the medieval period to a great extent organised in terms of individual, separate diseases, which is reflected in contemporary common phases such as 'a herb grows for every illness'.[91] This applied to medical advice about diet as well. Early seventeenth-century health guides were still frequently organised in a manner which allowed the reader to quickly find dietary advice to match their personal constitution. Books devoted sections or whole chapters to the 'sanguineans', 'cholericks', 'melancholicks', and 'flegmaticks' but also to children, the aged, and nursing women.[92] Other treatises provided long lists of aliments divided into sections about meat, vegetables, and drinks so that the

[91] Stolberg, *Experiencing Illness*, p. 25.

[92] For example, John Makluire, *The Buckler of bodilie health whereby health may bee defended and sicknesse repelled.* 2nd edition (Edinburgh: Printed by John Wreittoun, 1630), Early English Books Online; Maynwaring offered 'peculiar Courses', for different constitutions, ages, abilities, states and customs in his *The method and means of enjoying health.*

reader could look up a food item and determine whether or not it was appropriate to his or her constitution or circumstances, or whether it was the right season or even month to consume that particular aliment.[93] The focus of dietary advice was primarily on the qualities of the body and on the qualities of the food. By the early eighteenth century, however, this started to change, as dietary advice became increasingly specialised for, and organised according to, particular diseases and symptoms. The long lists of aliments and their qualities had all but disappeared from dietary advice by the 1720s, and although constitutions continued to feature in discussions of diet, these no longer took up the space that they had in the previous century.[94] Instead, dietary advice started to be structured around illnesses, such as Strother's 1725 chapter on diet, which argued that 'all Diet must be contrary to the Nature of Diseases' and provided specific dietary advice for readers suffering from fevers, inflammations, viscidity of the blood, kidney stones, fainting, atrophy, and a host of other diseases.[95] While Strother's work functioned more as a reference book for a host of illnesses, other physicians focussed on one particular disease or groups of diseases or symptoms they saw fit to discuss together.

The eighteenth century witnessed a rapid increase in publications specifying advice for particular diseases. As Stolberg argued, these were the result of developments not only in medical theory but also in medical practice and education. Medical students were required to publish a dissertation on a specific topic in order to obtain a doctorate, which at least in part explains the growing number of publications on all sorts of diseases.[96] These disease-specific treatises continued to emphasise the importance of the right management of the non-naturals, especially diet. This was particularly true for discussions about morbid corpulency, the state of fatness which was considered to be an illness, rather than a constitution. The first step to curing and preventing corpulency, physicians argued, was to remove all aliment of a rich or oily nature from one's diet.[97] The diet of the corpulent was to be of the 'less nutritious sort', mostly consisting of

[93] For example, Vaughan, *Approved directions for health*, and Muffett, *Healths Improvement*; Thomas Tryon organised diets per season in his *Way to Health*.

[94] Certain chemically inclined physicians did provide long lists of aliments and their chemical properties, such as Forster, *A Treatise of the various kinds and qualities of foods*.

[95] Strother, *An essay on sickness and health*.

[96] Stolberg, *Abhorreas pinguedinem*, 374.

[97] Malcolm Flemyng, *A discourse on the nature, causes and cure of corpulency* (London: Printed for L. Davis and C. Reymers, 1760), 6, Eighteenth Century Collections Online.

food which was 'lean and plain', and vegetable, rather than animal, as the former contained less oil and was more readily excreted.[98] In doing so, physicians dared to be a lot more specific in their dietary rules, forgoing any of the usual adjustments and exemptions for age or constitution frequently found in older and more general health guides. As a result, the diets designed for sufferers of specific diseases became increasingly generalised, with no explicit distinction between diets for old or young fat bodies, male or female fat bodies, or 'sanguineous' or 'cholerick' fat bodies. This is not to say that these physicians did not consider constitutions at all in discussions of corpulency. A naturally weak constitution was sometimes considered to be a cause, or at least a contributing factor, to the development of this disease. Constitutions were believed to affect the prognosis as well: Short noted that corpulency caused different indispositions in different constitutions: he believed gout, arthritis, and dropsy to be particular to fat persons of a cold constitution, for example.[99] Regardless, neither Short nor Flemyng adjusted for constitutions in their dietary advice.

Cheyne and Cocchi's treatises did not seek to cure or prevent one particular illness but were aimed at a wider group of people whom they believed to have certain dietary needs in common. Cheyne's 1720 *Essay on Health and Long Life* addressed 'the Sickly and the Aged, the Studious and the Sedentary, Persons of weak Nerves, and the Gentlemen of the learned Professions'.[100] His *Essay on Regimen* was written to relieve 'the Scorbutic, Gouty, Consumptive Valetudinarian-low-livers', and those 'whose vocation and occupation is sedentary and studious'.[101] Cocchi did not explicitly address his treatise to a particular audience, but he did suggest that the original intention of the Pythagorean diet (named after the diet followed by ancient Greek Philosopher Pythagoras of Samos) had been 'to cure Diseases and Corpulency, as well as gross Habits', and Cocchi himself devoted some pages to the curative effects of the Pythagorean diet on cases of the gout, scurvy, and nervous diseases.[102] Considering the general consensus was that diet ought to be specific

[98] Thomas Short, *A discourse concerning the causes and effects of corpulency: together with the methods for its prevention and cure* (London: Printed for J. Roberts, 1727), 72, Eighteenth Century Collections Online; Flemyng, *A discourse on the nature, causes and cure of corpulency*, 12.

[99] Short, *A discourse concerning the causes and effects of corpulency*, 60 and 63.

[100] Cheyne, *An Essay on health and long life*, xiv.

[101] Cheyne, *An essay on regimen*, xxx.

[102] Cocchi, *The Pythagorean diet*, 34.

according to one's particular disease, it would seem that Cheyne and Cocchi's method of grouping all these health complaints was directly at odds with their colleagues' advice. Cheyne had argued that most chronic illnesses shared a common cause and, consequently, also a common cure. 'Repletion' was a term used to describe an excess of matter in the body and was, as far as Cheyne was concerned, the cause not only of most illnesses but of all infirmities related to old age and the primary reason behind Englishmen's short lifespan.[103] Such an excess could result from overeating or issues with the body's ability to excrete sufficiently, or a combination of the two. Cheyne's choice of 'light' vegetable food, milk, and grains was selected due to their perceived effects on the circulation of the fluids. These foods were quickly digested, added but little bulk to the body, and were largely and easily excreted. Their aqueous nature diluted the fluid and helped 'flush out' any matter which was lingering and causing blockages. Such a diet of 'the lightest and the least', Cheyne argued, 'could supply the place of any specific regimen of diet'.[104]

Perhaps Cheyne was of the opinion that curing repletion thought diet warranted prescribing his patients a diet consisting of foods which were food aversions, or potential unwanted weight loss or loss of strength that a patient may suffer by eating only food which was 'light' and 'perspirable'. Cheyne admitted that this specific regimen was not sufficiently nourishing to be turned into 'muscular Flesh'.[105] He also believed that a total abstinence from meat could weaken the digestive powers of the body as the stomach and bowels were given very little to exercise themselves on.[106] Instead, these foods were more 'transitionary, fugitive and of prompt Secretion' and would 'perform all the animal Functions more readily and pleasantly, with fewer Resistances and less Labour' of the digestive organs.[107] Vegetable food was not a 'superfood' exactly: the benefit of the dietary course lay strictly in its ability to cure repletion. Of course, since Cheyne identified repletion as the primary cause of nearly all chronic illnesses, his light diet was to be presented as the de facto choice for a very large group of people which included not only the ill but also the sedentary, and all who had passed the age of 50. For Cheyne, it seems, the retention from which he believed these groups of people to be suffering

[103] Cheyne, *An Essay of health and long life*, 35–36.
[104] Cheyne, *An essay on regimen*, Lxi
[105] Cheyne, *The natural method*, 56.
[106] George Cheyne, *Observations concerning the Nature and due Method of treating the Gout* (London, printed for G. Strahan, 1720), 13–14.
[107] Cheyne, *The natural method*, 56.

overrode other factors which might have distinguished them, such as their sex, their constitutions, and the climate in which they normally resided.

Conclusion

Eighteenth-century dietary advice literature was a field of both tradition and innovation. While many aspects of Galenic medicine continued to form part of the dietary advice of this period, new ideas about the physiology of the body and the nature of food allowed physicians to experiment with new arguments and approaches. The iatromechanical school of thought was especially influential in early eighteenth-century medical circles and introduced a number of new theories and terms to dietary literature of this period, especially relating to the solid parts of the body. Mechanically inclined physicians suggested that these solid parts played a much more active role in the digestion and circulation of the alimentary fluids than previously thought. The body was not a sack of fluid but a solid hydraulic machine consisting of pipes and tubes through which the fluids were propelled forth. Consequently, dietary advice turned to preventing and curing symptoms and illnesses they believed to be caused primarily by weakened and blocked vessels. Preventing blockages was not simply a matter of avoiding excessive eating—choosing the right kind of aliment or food was just as important. Aliment was understood to be made up of different particles which were broken down over the course of the digestive process, and these particles had to travel through a vast network of tubes and openings functioning much like filters. The further the aliment travelled along the digestive process, the narrower these passages became. Physicians such as Cheyne argued that food needed to consist of particles which were less cohesive and more easily broken down into the finest matter so that it could ultimately be cast out of the body through insensible perspiration. Particles which remained too large to pass through the narrowest passages would be retained within the body, fill up those passages, and block new aliment from being digested fully.

The most easily digested aliments or foodstuffs, Cheyne and Cocchi argued, were milk and vegetables, and so this new focus on the digestibility of foodstuffs heralded in a period of intense debate about the benefits and limitations of diets which were low on, or completely free from, animal flesh. The trend towards vegetarianism was not the only consequence of mechanical approaches to diet, however.

Courses of diet consisting primarily of milk and vegetable foods were argued to be suitable for the elderly, the sedentary, and all those suffering from any kind of chronic illness. Cheyne even went so far as to argue that those who had a natural aversion to milk ought to consume it regardless of this aversion—advice which was all but unheard of in this period. This new focus on the solids of the body, so strong especially among mechanically inclined physicians, seems therefore to have opened up the possibility of talking more generally about rules of diet. The idea that digestibility was key to a healthy diet appears to have contributed to early eighteenth-century attempts at devising dietary rules applicable to large sections of the population, rather than just for specific individuals. Following the republication of Santorio's *Medicina statica* in several European languages, including English, the early eighteenth-century medical field witnessed a flurry of experiments conducted to discover how much the body excreted and consequently needed to ingest to make up for that daily loss. Especially among iatromechanists, the focus of dietary advice shifted from balancing to replenishing the body. What followed was an increased interest in managing dietary quantities rather than qualities, although concern with the latter did not disappear. Instead, physicians adjusted dietary advice in line with iatromechanical theory by changing which qualities they looked for in aliment. Through their experiments, Santorio, Keill, and Robinson had found that certain kinds of food added more to the body while other kinds were mostly excreted through sensible and especially insensible evacuations. These new qualities, often referred to as 'nourishing' or 'perspirable', entered the early eighteenth-century dietary vocabulary and were soon adopted by Cheyne and Cocchi, among others, who, in their efforts to cure people of gout, retention, and corpulency, constructed dietary courses consisting primarily of aliment which had a 'perspirable' quality to them. These kinds of food were not universally wholesome, they admitted, but they nevertheless advised to a relatively broad readership that such aliment was to be preferred over foods of a more nourishing nature and that simply the pursuit of the least and the lightest could replace any 'specific' diet. This period thus witnessed a unique shift where some physicians suggested courses of diet focussed on just a few qualities found in food, rather than a whole range of different qualities, and consequently were advised to a much wider group of people than had been possible hitherto. This shift did not take place over the whole genre of health advice. Many physicians still argued that dietary rules needed to be bespoke in nature in order for them to be safe and effective, and earlier ideas about the importance of constitution existed alongside the newer attention on the digestibility of certain food.

Another part of this development was that dietary advice, like other medical advice, became increasingly centred around diseases in the early eighteenth century. Treatises which set out to cure or manage particular illnesses or symptoms started to appear in increasing numbers and provided the sufferers of these illnesses with courses of diet specific to them. Some of the more general health guides, intended for a broader audience, replaced their long lists of ingredients and their qualities with lists of symptoms matched with aliments which had qualities that were contrary to these symptoms. Although the physicians who wrote these works had not eliminated constitutions from their discussions, their dietary advice seem to have become increasingly focussed on curing and managing distempers believed to be shared by many individuals. Cheyne in particular offered dietary advice which he believed to be suitable for the aged, the enfeebled, the sedentary, and the nervous, on the basis that these people all suffered from impaired digestion and would all benefit from food which was easily digested. Such a dietary regimen, of the 'least and the lightest', was far more general than that of the diets recommended by chemically inclined physicians, who believed an effectual diet needed to either maintain the blood's equilibrium of salts, alkali and acids, or be contrary in nature to particular diseases, of which there were a vast variety. Finally, it is also possible that this development was steered by more practical reasons, such as Cheyne seeking to expand his potential clientele by offering dietary advice which would have been generally beneficial and inoffensive for most people. The period in question did witness a rapid expansion of all printed literature, and it is not unlikely Cheyne sought to take advantage by appealing to as broad a readership as possible.

Early eighteen-century dietary literature therefore did witness some significant changes in what food choices physicians recommended and more importantly, how these recommendations were justified. The more extreme cases, such as the texts written by Cheyne and Cocchi, did not reflect the period's dietetic literature as a whole and were in fact met with considerable criticism. Most dietary treatises championed bespoke advice long after the peak of their success. This however should not lead us to conclude that these changes were insignificant. Their efforts to provide dietary advice to large groups of people rather than individuals deserve our attention especially *because* such an approach was unusual at the time. An investigation into dietary practices which seem ahead of their time reminds us that many practices of the past were the result of unique historical circumstances and often did not neatly follow a linear pattern. Dietary advice would eventually come to be applied to large groups of society with only minimal adjustments made for factors such as age, sex, and climate. Significant efforts to standardise diet and generalise dietary rules were made again in the late eighteenth century and eventually replaced the traditional advice in the early nineteenth century. As will be shown in the next chapter, these later efforts were supported with theories and motivations vastly different from those expressed by Cheyne and Cocchi.

CHAPTER 3

From Appetite to Measure? Food Quantities in British Dietary Advice

As we have seen in the previous chapter, food qualities remained an important part of food advice over the course of the seventeenth and eighteenth centuries. Yet it was neither the only nor indeed even most important factor to a healthful diet, physicians warned. 'I have said', Bath physician Cheyne argued in his *Essay of Regimen*, 'that the Quantity of Aliment is of much greater Consequence in general, than the Quality, for the Preservation of Life, Health and Serenity'.[1] Taking in too large a quantity of food could lead to a number of diseases, physicians believed, ranging from indigestion to corpulency.[2] Physicians argued that even excessive consumption of 'wholesome food' could lead to ill health. Concerns with eating the right quantities seemed to increase over the course of the seventeenth and eighteenth centuries. Any discussion on early modern dietary advice must therefore include a discussion of dietary quantities as well as qualities.

When and, more importantly, why dietary quantities became an important feature of health advice has been a central question of several

[1] George Cheyne, *An essay on regimen* (London: Printed for C. Rivington, 1740), xxvii, Eighteenth Century Collections Online.

[2] Thomas Short, *A discourse concerning the causes and effects of corpulency: together with the methods for its prevention and cure* (London: Printed for J. Roberts, 1727), 10, Eighteenth Century Collections Online.

M. Wübbels, *Quantification and Measurement in British Health Advice, c. 1670–1830*, Palgrave Studies in Medieval and Early Modern Medicine,
https://doi.org/10.1007/978-3-032-32760-4_3

historical inquiries into dietary literature and practices of the past. Jane O'Hara-May, one of the first to thoroughly investigate past attempts to measure man's dietary needs, recognised that the question of how much food a man should eat had been part of dietary advice since antiquity but argues that it was not until the mid-nineteenth century that 'a workable quantitative basis for calculating the food requirement of man was established'.[3] O'Hara-May argued that such a quantitative basis had not been possible under a humoural medical system, which ascribed humoural qualities to food as well as to bodies, all of which varied from foodstuff to foodstuff, body to body, and interacted with each other in complicated ways. This humoural system made it impossible to set dietary rules, whether in relation to dietary qualities or quantities, which were applicable to all people in general. Dietary advice from before the eighteenth century was therefore still very much focussed on the specific dietary needs of the individual, rather than 'the body' in general. In the nineteenth century, however, scientists began to recognise that the body and foodstuffs were composed of the same basic chemical elements, which allowed scientists to be more specific about the dietary needs of the body.[4] O'Hara-May argues that it was especially the work of Dutch physician Jacob Molleschott (1822–1893), published in the middle of the century in *Wiener Medizinische Wochenschrifte* (14 May 1859), which provided the first step towards acceptance of a quantified dietary standard.[5]

As we have seen in the previous chapter, the eighteenth century has long been seen as a pivotal point in the shift in focus in health literature from the health of the individual to the health of society. Maria Pia Donato argues that while at the beginning of the eighteenth century, healthy living was still a matter of managing the non-naturals of the individual, literature from the late eighteenth century was noticeably more concerned with health advice designed to protect and preserve public health.[6] Similar to O'Hara-May, Donato too attributes such a shift from individual to collective health to the decline of the doctrine of the humoural system. However,

[3] Jane O'Hara-May, 'Measuring Man's Needs', *Journal of the History of Biology*, Vol. 4, No 2 (Autumn, 1971), 249–273, citation taken from p. 249.

[4] Ibid., 250.

[5] Ibid., 273.

[6] Maria Pia Donato, "The afterlife of the non-naturals in early eighteenth-century Hippocratism: from the healthy individual to the healthy population," in *Conserving Health in Early Modern Europe: Bodies and environments in Italy and England*, edited by Sandra Cavallo and Tessa Storey (Manchester: Manchester University Press 2017), 174.

Donato argues that the emergence of an early form of environmental medicine with a focus on controlling matters of air, water, and food to prevent the outbreak of diseases in the wake of concerns about birth rates and population decline, also played a part in this shift.[7]

Rebecca Earle points to late-eighteenth-century statesmen concerned with the wealth and power of nations, who argued that a healthy population of workers needed nourishing food suitable for the labouring body.[8] It was with these concerns in mind, Earle argues, that eighteenth-century theorists tried to ascertain how much food an individual, and the nation, required. Emma Spary furthered this debate concerning the shift from individual to collective health problems.[9] Similarly to Earle, she too suggests that medical attempts to define and quantify a healthy diet should be seen in light of matters of statecraft.[10] 'A precondition for accomplishing the extension of statecraft to the body was the establishment of a universal model of dietary needs applicable to all', she argues.[11] These attempts started with experiments in hospitals, workhouses, and prisons and were designed to optimise production while minimising the cost of feeding the working population. These trials were then used as evidence of the quantity of nourishment required to sustain life.[12] In *Imagining Poverty*, Sandra Sherman argued that eighteenth-century scientific literature purposefully represented the poor as a uniform cohort.[13] 'The poor were abstracted, homogenized', put at a distance by an 'avalanche of printed numbers', Sherman notes, that replaced individual needs with statistics, ratios, and accounts.[14] These accounts were produced in an effort to limit the expenses of 'disciplinary spaces' such as workhouses whilst maintaining the productivity of the workers in these spaces.[15] If, as Sherman argues, these accounts took but little notice of individual circumstances, it may be this development in particular which contributed to a generalisation of dietary advice.

[7] Ibid., 159.

[8] Rebecca Earle, "The Political Economy of Nutrition in the Eighteenth century," *Past and Present*, 242, 1 (Feb. 2019), 83.

[9] Emma Spary, "Food," in *A Cultural History of Medicine: In the Enlightenment* ed. Lisa Wynne Smith (London: Bloomsbury 2021), 29–50.

[10] Ibid., 46.

[11] Ibid., 47.

[12] Ibid., 47–48.

[13] Sandra Sherman, *Imagining Poverty: Quantification and the Decline of Paternalism* (Ohio State University 2001), 3.

[14] Ibid.

[15] Ibid.

After all, as Spary argued, these workhouse trials would eventually be used as evidence for more general conclusions about the dietary needs of people and thus contribute to the development of programmes for the management of public diet.[16] There appears to be a general consensus then that it was sometime over the course of the late eighteenth and early nineteenth centuries, that attempts to quantify dietary needs and policies to push for collective health guidance met and resulted in experiments designed to measure how much food groups of people needed on a daily basis.

However, as an ever-expanding scholarship on the history of quantification in medicine has amply demonstrated, interest in determining how much food the human body needs to function predates the late eighteenth century.[17] One of the first known attempts to quantify one's dietary requirements was undertaken by Paduan nobleman Alvise Cornaro, often Italianised to Luigi Cornaro (1475–1566). Cornaro followed a strict regimen to recover from a lifetime of debauchery, which had caused him many ailments. He experimented with his diet by trying to ascertain how much (or rather, how little) food he needed to sustain himself and live for as long as possible. After much trial and error, Cornaro determined that he was most healthful under a diet consisting of 12 ounces of food exactly weighed ('pesusse oncie dodecie alla solite').[18] In early modern Italy (and indeed also in early modern Britain) units of measurements were not yet standardised and varied from place to place, but Laura Madella puts this amount at around 270–470 grams of food and around 320–550 grams of liquid in a day if Cornaro's account was truthful.[19] Cornaro published the

[16] Spary "Food", 47–48.

[17] See especially Jonathan Barry and Fabrizio Bigotti (eds.), *Santorio Santori and the Emergence of Quantified Medicine, 1614–1790: Corpulscularianism, Technology and Experimentation* (Cham: Springer International Publishing AG, 2022), and Simone Guidi and Joaquim Braga (eds.), *The Quantification of Life and Health from the Sixteenth to the Nineteenth Century: Intersections of Medicine and Philosophy* (Palgrave Studies in Medieval and Early Modern Medicine, 2023).

[18] Luigi Cornaro, *Discorsi della vita sobria … ne' quali con l'esempio di se stesso dimostra con quai mezi possa l'huomo conservarsi sano insin'all'ultima vecchiezza* (1699), 19.

[19] As Madella explains, scholars have found discrepancies in Cornaro's claims, especially regarding how old he purported to be. This was relevant, as the success of *Discorsi della vita sobria* was often explicitly linked to Cornaro's claims that his lifestyle had lengthened his life. Laura Madella, "The More the Year the Less the Food: Alvise Cornaro on *The Sober Life* (1558)", Simone Guidi, and Joaquim Braga (eds.), *The Quantification of life and health from the sixteenth to the Nineteenth century: Intersections of Medicine and Philosophy,* in Palgrave

details of his regimen in *Discorsi della vita sobria* (Padua, 1565), which was first translated into English in 1591 and became particularly popular in Britain, where more than a hundred editions were printed.[20]

The first static experiment designed to ascertain how much food the human body needed was undertaken by Santorio Santori (1561–1636) in the late sixteenth century. For a period of circa 30 years, the Paduan physician weighed himself and others to measure how much the body excreted in a 24-hour period and how much a person had to eat to make up for that loss.[21] The results of this weighing experiment were published in Santorio's most famous work, the *Ars de statica medicina* (Venice, 1614). Santorio's experiment was repeated over the course of the eighteenth century, by physicians such as the Irish George Rye (between 1721 and 1722) and Bryan Robinson (between 1721 and 1745), Scottish physician Francis Home (in 1751), and British physicians James Keill (in 1708) and William Stark (in 1769).

A rich scholarship exists on the quantification of medicine, which is generally said to have started in the late seventeenth century.[22] However, the effects of quantification on late seventeenth- and eighteenth-century popular health literature have not yet been studied in depth, despite the ongoing popularity of this genre. This is an oversight this chapter seeks to rectify. This chapter will ask whether dietary advice literature in this period witnessed a shift in focus from qualitative methods to quantitative methods for determining how much a person should eat. The chapter will also

Studies in Medieval and Early Modern Medicine, Palgrave Studies in Medieval and Early Modern Medicine, (London, Palgrave Macmillan Cham, 2023), https://doi.org/10.1007/978-3-031-15725-7, 74–76.

[20] Marisa Milani, "Introduction to Cornaro" in *Writings on the Sober Life: the Art and Grace of Living Long*, ed. and trans. by Hiroko Fudemoto (Toronto: University Press, 2014), 38.

[21] For details concerning the dates and methods of Santorio's experiment, see the introduction in Barry and Bigotti, *Santorio Santori and the Emergence of Quantified Medicine*, esp. p 28–32, and 38.

[22] Fabrizio Bigotti, 'The Weight of the Air: Santorio's Thermometers and the Early History of Medical Quantification Reconsidered', *Early Modern Studies*, Vol. 7, Issue 1 (2018), 73–103; Andrea A. Rusnock, 'The Weight of Evidence and the Burden of Authority: Case Histories, Medical Statistics and Smallpox Inoculation', *Medicine in the Enlightenment*, vol 29 (Feb 1995); Andrea A. Rusnock, Vital Accounts: Quantifying Health and Population in Eighteenth-Century England and France (Cambridge, 2002); Frangsmyr, Tore, J.L. Heilbron, and Robin E. Rider (eds) *The Quantifying Spirit in the Eighteenth Century* (Berkeley, University of California Press, 1990).

assess to what extent an increased focus on quantitative methods for measuring diet enabled physicians to set dietary standards for groups of people (grouped by gender, age, or activity levels) rather than just for individuals. My initial inquiry started with the hypothesis that early metabolic experiments such as the one undertaken by Santorio could have contributed to a shift away from qualitative methods for determining dietary intake such as adhering to one's appetite towards more quantitative methods such as through measuring food or the body. I theorised that if there was such a shift, then this may have also led to more dietary rules aimed at a wider population, rather than at individuals. However, my hypothesis was not borne out by the research, and I will conversely demonstrate that while the period did witness considerable interest from both physicians and patients in quantifiable methods of managing diet, these did not come to replace qualitative methods. I will also show that these methods did not cause physicians to design or support rules of health which applied to all or most people. In fact, health advice remained highly personalised throughout the period under consideration.

Appetite and Moderation

'The Rule for eating to every body, is a natural Appetite', Scottish physician James Keill (1673–1719) argued in his aphorisms.[23] It was through adhering to one's appetite that the just quantity of food was taken in, without any need for weighing or measuring that quantity, 'for Nature never requires more or less than is convenient, and the Appetite is proportionable to the natural Evacuations', he assured his readers.[24] He was not alone in this conviction. 'The natural Appetite is certainly the best Guide, both in the Quantities and Qualities of the Meats and Drinks to be taken in', John Quincy (unknown to 1722) commented.[25] 'But', he added as a warning, 'persons must be careful to distinguish a natural Appetite from one that is vitiated and debauched by excess and over Indulgence'.[26] And herein lay exactly the issue. Over the course of the seventeenth and

[23] James Keill, in *Medicina Statica: being the aphorisms of Sanctorius, translated into English, with large explanations. To which is added, Dr. Keill's Medicina Statica Britannica... By John Quincy.* 5th edition (London: Printed for T. Longman, 1737), 263. Eighteenth Century Collections Online.

[24] Ibid.

[25] Quincy commenting on James Keill, in *Medicina Statica* (version 1737), 262–263.

[26] Ibid.

eighteenth centuries, physicians expressed the concern that the influx of new foodstuffs had altered the appetite of the British people and not for the better. Bath physician George Cheyne (1671–1743) lamented:

> Since our Wealth has increased, and our Navigation has been extended, we have ransacked all the parts of the Globe to bring together its whole stock of Materials for riot, luxury, and to provoke excess. The tables of the rich and the great (and indeed of all Ranks who can afford it) are furnished with provisions of delicacy, number, and plenty, sufficient to provoke, and even gorge, the most large and voluptuous Appetite.[27]

It was not only the influx of new produce that tempted the appetites of the well-to-do. New cooking styles, especially those coming from France, were also to blame as far as Cheyne was concerned:

> The ingenious mixing and compounding of Sauces with foreign Spices and provocative, and contrived, not only to rouse a sickly Appetite to received [sic] the unnatural Load, but to render a natural good one incapable of knowing when it has enough. Since French cookery has been in such repute in England, and has been improved from Spain, Italy, Turkey and every other country that has any things remarkably delicious, high, or savoury in Food; since eastern Pickles and Sauces have been brought to embellish our continual Feasts. Dressing, which was design'd to assist the labour of Digestion, as it is now manag'd, not only counter-acts that design, but is become the most difficult, ingenious, and, at the same time, one of the most profitable Trades.[28]

While warnings of the dangers of overeating were hardly a new feature in health guides, the influx of new ingredients and recipes seemed to have led to an increased focus on how much people ate and how much they ought to eat. At the centre of this debate was the question of how people should determine the right quantity of food for their bodies.

The concept of moderation in food and drink was ubiquitous in early modern health advice. The concept was not well-defined, however, as Anita Guerrini has recently shown and the practice of moderation varied

[27] George Cheyne, *The English malady: or, a treatise of nervous diseases of all kinds* (London: Printed for G. Strahan, 1733), 49–50, Eighteenth Century Collections Online.

[28] Ibid., 52.

widely.[29] The most commonly found rules for eating moderately were qualitative and subjective in nature. For example, the Flemish Jesuit Leonard Lessius (1554–1623), whose essay *Hygiasticon* was frequently published together with Cornaro's *Discorsi della vita sobria*, proposed seven rules for discovering the 'right Measure' in diet, most of which involved observing the body for signs of having exceeded the limit. Lessius's first rule advised that if a man would take in so much food and drink as to 'render him incapable or unfit for his mental Employments such as Prayer, Meditation, speculative Studies and the like', he had exceeded the 'measure which he ought to keep'.[30] Other signs of having exceeded the limits of moderation included 'Dulness, heaviness of Disposition' or an 'Oppression of the Stomach'.[31] To prevent overeating, Lessius recommended that people 'forbear Variety of Dishes, and the luxurious Artfulness of Cooking'.[32] Such 'tampering' with food caused an appetite beyond the natural cravings of hunger and thirst, Lessius warned.

In the eighteenth century, many physicians echoed Lessius's advice and continued to refer to qualitative methods for measuring the right quantity of food. Alongside this tradition, however, we find increasing interest in quantitative methods for finding out how much food a person should consume daily. Under other circumstances, hunger and thirst *would* be 'the surest Guides both as to the Quantity and Quality of what we either eat or drink', English physician Jeremiah Wainewright explained in 1707, but many had debauched their appetites by 'Irregularity and excess', to the extent that they no longer possessed the skill to determine the quantities they were taking into their stomachs.[33] The exotic and rich aliments offered temptation that had not been present in the British diet before, physicians argued, and drew those who indulged in them past the point of fullness. The British elite was losing their ability to gauge when enough

[29] Anita Guerrini, 'The impossible ideal of moderation', in James Kennaway and Rina Knoeff (eds), *Lifestyle and Medicine in the Enlightenment: the six non-naturals in the long eighteenth century* (Abingdon: Routledge, 2020), 86.

[30] Leonard Lessius, *Hygiasticon: Or, a Treatise of the Means of Health and* Long *Life, Written originally in Latin by Leonard Lessius, Now rendered into English by T. Smith, Whereunto is annexed, Cornaro's* Treatise *of the benefits of a Sober Life* (London: Printed for the Author and sold by Charles Hitch, 1742), 18.

[31] Ibid., 22.

[32] Ibid., 36.

[33] Jeremiah Wainewright, *A mechanical account of the non-naturals: being a brief explication of the changes made in humane bodies, by air, diet &c* (London: Printed for Ralph Smith, 1707), 150–151, Eighteenth Century Collections Online.

had been enough: as a result of new imported goods and cooking fashions, their natural appetite had now well and truly been spoilt.

As, in this climate, appetite could no longer be trusted to guide eaters in determining the right quantity of aliment, physicians' rules were more needed than ever, physicians argued. 'Pleasure blinds us that we cannot discern when we are come to the due measure we ought to hold', Lessius explained, 'but hides the bound-marks thereof, to draw us past them'.[34] To prevent this, Lessius continued, certain rules had to be produced, 'whereby it may be clearly and certainly found out'.[35] Readers of seventeenth- and eighteenth-century health guides found themselves advised to abandon trust in their appetites in favour of following rules independent of their sensations.[36] Of course, dietary rules had been a mainstay of health advice since Galen and Hippocrates, and the idea that they had to be followed in order to obtain and preserve health was a strong theme throughout early modern medical literature. '[T]hey that desire to live long … [l]et them conform and be obedient to the Hygiastick Laws, and wholesome Rules hereafter prescribed', Maynwaring advised.[37] Contrarily, 'they that live carelessly and irregularly, contemning Physical Rules, as unnecessary Observations', would have to resort to remedies.[38]

Though most physicians continued to advise readers to monitor their bodies and bodily functions for signs of having exceeded the limit, a handful of writers, including the famous Cornaro and Lessius, attempted to quantify the right amount of daily food. Cornaro described his 'usual

[34] Leonardus Lessius, in *The Temperate Man, or the Right Way of Preserving Life and Health … In Three Treatises, the first written by the learned Leonardus Lessius, The second by Lodowick Cornaro, a Noble Gentleman of Venice, the Third by a famous Italian*. (London, printed by J.R. for John Starkey, 1678), 13.

[35] Ibid.

[36] Cheyne, *An Essay of health and long life* (London: Printed for George Strahan, 1724), 39, Eighteenth Century Collections Online.

[37] Everard Maynwaring, *The method and means of enjoying health, vigour and long life adapting peculiar courses for different constitutions, ages, abilities, valetudinary states, individual proprieties, habituated customs, and passions of mind* (London: Printed by J.M. for Dorman Newman, 1683), Early English Books Online, in the section, 'The scope of this work', unpaginated. 'Hygiastic Laws' refers to the rules set out in Lessius's *Hygiasticon*, the English version of which was published just five years before Maynwaring published his *Methods and Means*.

[38] Maynwaring, *The method and means of enjoying health*, in 'The scope of this work', unpaginated.

measure' of food as 'twelve Ounces exactly weighed', '*per diem*'.[39] This quantity was measured for himself and was not the quantity of food he recommended for others. Indeed, Cornaro emphasised strongly that every man was to discover the right quantity for himself and that no one rule could possibly apply to all or even most people. Cornaro advised: 'Every Man, by long Observation and various Experiment, may attain to Perfect Knowledge of the Qualities of his own Nature', and 'learn precisely the Quantity and Quality of Meat and Drink, that are lightest upon his Stomach; which cannot certainly be so well understood by another, as by himself'.[40] Lessius was more willing than Cornaro to suggest that Cornaro's dietary quantity was suitable to more people than just Cornaro. Lessius advised that:

> While the exact Proportion cannot be precisely adjusted for each person considering the Various Differences of Mens Age, Strength and Dispositions ... Yet (generally speaking) for those, who are much advanced in Years, and for those also, who are of Weak Complexions, twelve or fourteen Ounces of Food a-day are judged sufficient, and are what ought not to be exceeded.[41]

Attempts to quantify moderation continued in the eighteenth century. Bath physician George Cheyne offered 'to determine the Quantity of Food', in his *Essay of Health and Long Life* (1725) and put this quantity at '8 Ounces of Flesh Meat, 12 of Bread, or vegetable Food, and about a Pint of Wine, or other generous Liquor in 24 hours'.[42] Scottish physician James Keill came to the amount of four pounds of daily food, which medical writer John Quincy agreed was 'the Quantity convenient to most People'.[43] It is important to note here that at this point in time, weights and measures were not yet standardised and varied widely from place to place, even within Britain.[44] It can therefore be difficult to establish to what extent physicians followed each other's recommendations or diverged

[39] Cornaro, 'of the Benefits of a Sober Life', annexed to *Hygiasticon*, 19–20.

[40] Ibid., 22–23.

[41] Lessius, *Hygiasticon*, 27.

[42] Cheyne, *An essay of Health and Long Life*, 33–34.

[43] Keill and Quincy, both in *Medicina Statica Brittanica*, annexed to *Medicina Statica* (1720 version), 336.

[44] For a particularly clear explanation the variance of measurements in this period, see Guerrini, 'The impossible ideal of moderation', 100–101; and Ronald Zupko, *Revolution in Measurement* (Philadelphia: American Philosophy Society, 1990), especially 1–7 and 28–31.

from it. What is clear, however, is that these physicians thought it worthwhile to try to quantify moderation and to engage with other physicians' attempts to measure how much food a person should consume daily.

Quantifying Health and the Body: Santorio in the Eighteenth Century

The influx of exotic foods and new cooking fashions were not the only causes for the interest in quantitative methods for managing food intake. Seventeenth- and eighteenth-century physicians of the iatromechanical school of thought favoured quantitative methods for determining how the body functioned. Iatromechanism (sometimes also called iatrophysics or iatromathematics) was a school of thought rising in popularity in the medical circles of the late seventeenth century, following the works of scientists and physicians such as William Harvey (1578–1657), René Descartes (1596–1650), Giovanni Alfonso Borelli (1608–1678), Marcello Malpighi (1628–1694), Antoni van Leeuwenhoek (1632–1723), and Archibald Pitcairne (1652–1713). Iatromechanism was a school of thought rather than one particular theory or set of ideas which shifted and developed over time and can therefore be difficult to summarise. One of the key aspects of this medical movement was the view of the body as a mechanical construction. Iatromechanists often likened the body and its digestive processes to the functioning of a hydraulic machine, where the fluids, consisting of minute particles, flowed through an ever-narrowing network of tubular vessels. Followers of this school of thought moved away from the Galenic focus on individuals and their unique dietary needs and instead set to measuring and explaining the digestive process, which they argued functioned essentially the same in all bodies. Experimentation through measurements was considered an effective method for establishing rules about the functioning of the human body and how to keep the body healthful.[45]

[45] This aspect of iatromechanism has recently enjoyed particular interest from scholars, resulting in several works of note. Jonathan Barry and Fabrizio Bigotti have done extensive research on Santorio's quantitative research methods and measuring instruments and the effects of these on the field of medicine more broadly, in Barry and Bigotti, *Santorio Santori and the Emergence of Quantified Medicine.* See also the more recent publication of Simone Guidi and Joaquim Braga (eds.), *The Quantification of Life and Health from the Sixteenth to the Nineteenth Century: Intersections of Medicine and Philosophy* (Palgrave Studies in Medieval and Early Modern Medicine, 2023). See also Lucia Dacome, "Living with the chair: private excreta, collective health and medical authority in the eighteenth century," *History of Science*,

Physicians of the iatromechanical school of thought in particular argued that measurement was more reliable for gauging what quantity of food was healthful for an individual than appetite was, as it did not carry the risk of alimentary needs being distorted by personal appetites. It was in this medical milieu that eighteenth-century physicians revisited the weighing experiment undertaken by Santorio in the late sixteenth century.

Santorio's theories and methods were appropriated rather than adopted by eighteenth-century physicians, who were keen to have Santorio's calculations support iatromechanical theories to which Santorio himself would not have subscribed.[46] Santorio did not yet know, for example, of the circulation of the blood, which Quincy, author of the many translations of *Medicina statica* published between 1712 and 1737, argued was dependent on the elasticity of the fibres of the body.[47] As such, Santorio's aphorisms came to be explained in iatromechanical terms and were also used to support iatromechanical theories. Quincy published Santorio's aphorisms with extensive comments and explanations to ensure that these sixteenth-century rules fit neatly into eighteenth-century medical theory. The debates and methods of the early eighteenth century were not the result of a natural progression from Santorio's work but rather a development unique to the medical culture at the time, especially among Newtonian circles, and potentially spurred on by contemporary concerns around changing dietary habits.

Although regulating food intake was not the primary objective of the weighing chair, Santorio considered it a useful and reliable method for this purpose.[48] This was achieved through the dropping of the seat of the chair once the weightwatcher had reached a certain weight, which resulted in the victuals becoming physically out of reach. As Dacome points out, in some versions of the famous image of Santorio's weighing chair, which

39 (2001), 467–500; Lucia Dacome "Balancing Acts", *Studies in History and Philosophy of Biological and Biomedical Sciences*, 43, (2012); Michael Stolberg, 'Abhorreas pinguedinem: Fat and obesity in early modern medicine (c. 1500–1750)', *Studies in History and Philosophy of Biological and Biomedical Sciences*, 43, 2 (2012), 370–378; and Alexander Pyrges, "'No Age Did Ever Afford More Instances of Corpulency'. Obesity as a Collective Condition and the Early Modern Medicalization of Girth", in Jill Burke, Holly Fletcher, Christine Ott (eds.), *Fat Bodies in Early Modern Europe* (Forthcoming), which all investigate quantitative methodology in relation to seventeenth- and eighteenth-century medical discussions about body weight and how the practice of body weighing spread outside of the medical circles.

[46] Lucia Dacome argues this as well, in "Living with the chair".

[47] Quincy, in *Medicina Statica* (version 1737), p. 176.

[48] This has been extensively discussed by Lucia Dacome, "Balancing Acts", 381.

accompanied many of the eighteenth-century editions of *Medicina statica*, the figure being weighed gestures to the now out-of-reach table, presumably to draw the reader's attention to this aspect of the weighing chair (see the cover art of this book for an example).[49] It is unclear whether this particular aspect of the weighing experiment enjoyed much popularity after Santorio. Dacome points out that Santorio considered methods for hiding the beam of the machine so that it would not look 'ridiculous' to visitors, which she suggests indicates that he envisioned the chair being used by a larger public than just medical men.[50] Moreover, at least one other physician devised a portable version of Santorio's chair and included the method of the ascending table, suggesting that others considered this to be a valuable aspect of the weighing experiment.[51] There is, however, no evidence beyond this that the ascending table was part of seventeenth- and eighteenth-century weighing practices. Experimenters such as Keill, Robinson, and Stark did not mention using a table, and the few sources we have of body weighing outside of physicians' experiments do not mention that this was part of the weighing process, all of which suggests that this aspect of Santorio's weighing practice did not catch on.

The method of determining food quantities through weighing experiments, however, certainly did. The work of Santorio, and the link it established between health and body weight, lived on in the medical circles of England long after the publication of *Medicina statica* in 1614. His statistical approach to maintaining the health of the body garnered significant interest from physicians in the late seventeenth century until the late eighteenth century, some of whom even re-tried the Santorian weighing experiment.[52] As recorded in Scottish physician James Mackenzie's *History of Health* (1757), the re-trials started with a French physician in 1668 in Dodart, France, and ended with William Stark's experiment, the results of which were published in his *Experiment on Diet*, in 1769.[53] The purpose behind these trials was generally to create aphorisms very similar to those of Santorio, except that the new calculations would account for the

[49] Ibid.

[50] Dacome, "Living with the chair", 476.

[51] Ibid.

[52] Ibid., 471–472.

[53] For a comprehensive overview of the experiments undertaken in this period, see James Mackenzie, *The history of health, and the Art of preserving it* (Edinburgh: Printed and sold by William Gordon, 1758), 260–264, Eighteenth Century Collections Online; or, Dacome, "Living with the chair", 481–487.

difference between Santorio's warm Italian climate and the cooler northern climates of England, Scotland, Ireland, Holland, and North America. In doing so, Dacome argued, these physicians lent further credit to weight-watching as a method of establishing rules about the body, especially in relation to the balancing of the *ingesta* (intake of food) and *excreta* (the sum of all the excretions of the body).[54] Alongside these re-trials, translations of Santorio's work started to appear from the late seventeenth century onwards, introducing the medical significance of weight to an audience beyond those who could read Latin. There is little doubt therefore that Santorio and his experiment enjoyed particular popularity in the early eighteenth century especially. Santorio, Keill stated in 1738, 'always will be applauded in every age of Learning, for having first discovered, and brought to perfection the way of measuring insensible Perspiration'. His aphorisms, Keill continued, were received by other physicians as 'universal Axioms'.[55]

However, the response to Santorio's work was not all praise. English essayist and politician Joseph Addison (1672–1719) criticised Santorio and his 'Mathematical chair' in a letter published in *The Spectator* in 1711.[56] Addison claimed to have received a letter from 'one of that sickly Tribe who are commonly known by the name of Valetudinarian' (a term popular in the eighteenth century to describe someone who was generally sickly), who related his experience of managing his diet by using a chair similar to Santorio's. The 'Valetudinarian' wrote:

> Having provided my self with this Chair, I used to Study, Eat, Drink and Sleep in it: insomuch that I may be said, for these last three Years, to have lived in a Pair of Scales. I compute my self, when I am in full Health, to be precisely Two hundred Weight, falling short of it about a Pound after a Day's Fast, and exceeding it as much after a very full Meal; so that it is my continual Employment to trim the Balance between these two Volatile Pounds in my Constitution. In my ordinary Meals I fetch my self up to Two hundred Weight and a half Pound; and if after having dined I find my self fall short of it, I drink just so much Small Beer, or eat such a quantity of Bread, as is sufficient to make me weight. I do not dine and sup by the Clock, but by my Chair; for when that informs me my Pound of Food is

[54] Dacome, "Living with the chair," 468.

[55] James Keill, *Essays on the several parts of the animal oeconomy*. London: Printed for George Strahan, 1738, Unpaginated Preface. Eighteenth Century Collections Online

[56] Joseph Addison, 'No. 25', *The Spectator*, 1, 25 (March 29, 1711), 92–95.

exhausted, I conclude my self to be hungry, and lay in another with all Diligence.[57]

This excerpt is especially vivid in its portrayal of how the subject of this letter replaces old dietary rules with a novel method by measuring and weighing himself to a healthful diet instead of adhering to physical cues of appetite. However, the letter was clearly meant to satirise Santorio's methods and was almost certainly written by Addison himself.[58] As such, the fictional Valetudinarian is no evidence of early eighteenth-century measuring practices but instead an imagined 'straw man' for Addison to point at in warning to those who would temper their health 'by Ounces and Scruples', instead of complying with 'natural Solicitations of Hunger and Thirst'.[59] Nevertheless, it is informative in that it portrays how contemporaries imagined that this method of managing dietary intake might affect daily life when taken to the extreme.

Addison was clear in his belief that the weighing experiment should not be carried beyond medical circles, and he was not alone in this conviction. In fact, while Santorio's epistemology received praise from those who would later carry out similar experiments, these later weightwatchers were explicit in their opinion that non-medical people should not follow their and Santorio's examples. In fact, the physicians who followed Santorio's example by conducting re-trials actively tried to dissuade readers from conducting experiments of their own. Rather, these physicians intended for the findings of their experiments, usually delivered in aphorisms, or tables, or both, to provide sufficient guidance as to how to manage the non-naturals. In an advertisement for Dutch physician Johannes de Gorter's *Treatise of Insensible Perspiration* (*Verhandeling van de onzigtbare doorwaasseming*), it was suggested that there were two methods for people to learn about insensible perspiration and the best methods for balancing these *excreta* with the right quantity of food. The first was by undertaking weighing trials. The second was by learning and applying the knowledge already provided in the aphorisms which had resulted from the weighing trials undertaken by

[57] Ibid., 93–94.

[58] Lucia Dacome and Fabiola Zurlini have come to the same conclusion that the letter was in fact written by Addison himself. See Dacome, 'Living with the chair', 479; and Fabiola Zurlini, 'The Uncertainty of Medicine: Readings and Reactions to Santorio Between Tradition and Reformation', in J. Barry and F. Bigotti (eds) Santorio Santori and the Emergence of Quantified Medicine, 1614–1790 (Palgrave Macmillan Camp, 2022), 111–112.

[59] Ibid., 95.

medical professionals.[60] The first was undoubtedly the most exact if properly executed, but the experimenters were explicit in their recommendations that readers of these treatises apply the second method to themselves. De Gorter purposely constructed his aphorisms so that anyone was able to judge how the excretions work in a body without needing the scales.[61]

De Gorter's experiment and findings were later summarised by Edinburgh physician James Mackenzie (1680–1761) and published in his *History of Health* (1758), along with summaries and comments on the experiments of Santorio, Keill, and Robinson.[62] Mackenzie, drawing from the aphorisms of all these experimenters, ultimately advised his readers that Santorio's experiment was 'a tedious method' and that De Gorter's recommendation that a man should eat and drink to satisfy his hunger and thirst, was 'the proper answer' to the question of how much aliment and drink was healthful. Most people would simply not be able to conduct similar experiments, even if they wanted to. There were many things 'related to the Animal Oeconomy' which 'may certainly be found out by weighing', Keill advised, 'but', he added, 'as they require a constant and certain way of living, a man of business cannot find leisure to pursue them with sufficient diligence'.[63] John Quincy, in his translation of Santorio's *Medicina statica*, agreed. He wrote in his letter to the readers:

> I am not at all unaware how severe some will be hereupon, in requiring how often they must weigh themselves, and whether they ought to drink and eat by the Ounce? To whom I have only this to say, that Sanctorius by the Balance, has already done enough to convince any serious person of the natural Discharges, and their Proportion to one another … from which … any Person may soon be a Judge of the present State of his Constitution, without going into a Pair of Scales.[64]

In fact, Quincy admitted to having deliberately omitted the 'Exactness of the Santorian Calculations', because, for all intents and purposes, it was

[60] "Verhandeling Van De Onzigtbare Doorwaasseming door Johannes De Gorter", *Maandelyke Uitreksels, of Boekzaal der Geleerde Waerelt.* 43, Chapter IV. T'Amsteldam: printed by Adriaan Wor and Erve Gerard onder de Linden (1736), Google Books, 321.

[61] Ibid., 321–322.

[62] James Mackenzie, *The history of health, and the Art of preserving it* (Edinburgh: Printed and sold by William Gordon, 1758), 257–264, Eighteenth Century Collections Online.

[63] Keill, *Essays on the several parts of the animal oeconomy*, unpaginated Preface.

[64] Quincy, *Medicina statica* (version 1737), vi.

sufficient to understand that there were such discharges, how such discharges could be influenced, and the consequences of them becoming disrupted.[65] 'Besides', Quincy added, 'were a Person to make Experiments with the Balance, it is not at all likely that they should exactly agree with Sanctorius's Accounts, both our Climate and Way of Living being so very different from his'.[66] The experiments were extremely difficult to replicate. Quincy pointed out that everything from the local temperature to non-natural factors could influence the results. As such, in his comment on Santorio's aphorism concerned with finding the right time for eating, Quincy advised that a 'Trial by Balance' would be 'too troublesome to put in practice' and that even 'if the Experiment should be one try'd, and the Times of Eating Settled accordingly, it is a great Chance but some Accident or other may make it convenient to alter them by the very next Day'.[67] Both the inconvenience of daily weighing and the many variables which could influence the weight of the body should be reason enough to convince any reader of the futility of conducting their own experiment, these physicians argued.

It is difficult to gauge whether Santorio had intended for body weighing to become the primary method for managing food intake. He did not deny the role of the embodied experience of the eater in determining how much food was the right amount of food. However, it does appear that Santorio ascribed a certain precision to the method of weighing which could not be found in qualitative methods. For example, the idea that the right quantity of food was that quantity which could be digested with ease was espoused by Santorio as well, though with the added caveat that easy digestion was to be measured by weighing:

> That Quantity of Food to everyone is most healthful, which without any Uneasiness can be perfectly digested: And that it is perfectly digested, may be known by the Sum of the Evacuations answering the Quantities taken in; which will appear by weighing.[68]

Santorio's assertion that it will appear by weighing, of course, does not indicate necessarily that other methods could not be as good, or at least, *good enough*, for determining whether the body had digested well or not.

[65] Ibid.
[66] Ibid.
[67] Ibid., 221.
[68] Santorio, *Medicina Statica* (version 1737), 187.

However, his argument that the state of digestion may be ***known*** by the sum of the evacuations suggests that for Santorio the quantitative methods presented a surer method of determining the right quantity of food. This notion that only the scales could offer any certainty about physiological processes such as digestion also shone through in other aphorisms. For example, Santorio argued that the 'most proper Time of Eating' had arrived when 'the Body comes to some healthful Standard, as it enjoyed the Day before, when empty'. 'But', he added, the exact timing of that equilibrium 'Apollo himself cannot find out, without the Balance'.[69] Thus, it appears that Santorio did consider the method imperative to knowing, with certainty and precision, how much a person should eat in order to remain healthful.

Mackenzie, in his *History of Health* (1757), provided his readers with a convenient summary of all the weighing experiments which had taken place up to the time of his writing, as well as some of the aphorisms resulting from these. The overview helps put into perspective how the weighing experiments were generally delivered to the eighteenth-century public and in what manner later experimenters diverged from Santorio. Addressing the question 'How shall a man fix upon the precise quantity of perspiration, which will secure him a permanent state of good health to old age?', Mackenzie provided answers from several experimenters, starting with Santorio, who had stated that it might be found by a weighing experiment.[70] This was quickly followed up with Keill's argument that 'the true rule of diet to every man, is his natural undepraved appetite' and that by this method he may know it 'without the trouble of weighing himself'.[71] Mackenzie then followed with De Gorter's aphorism, which, similarly to Keill, stated that if a healthy man 'eats and drinks as much as is sufficient to satisfy his hunger and thirst' and 'rises from the table without quite filling his stomach', 'his daily discharges will be equal to what he has taken in'.[72] Consequently, we can see that if Santorio had harboured any desire to make weighing experiments a popular method of measuring alimentary quantities, he did not succeed. His revered aphorisms were delivered to British readers with the added message that they should trust the

[69] Ibid., 203.
[70] Mackenzie, *The history of health*, 268.
[71] Ibid., 269.
[72] De Gorter, cited in Mackenzie, *The history of health*, 269.

aphorisms which came forth from the experiments to guide them to health, rather than conduct weighing experiments of their own.

In eighteenth-century dietary advice, the embodied experience continued to play an important part in determining the correct quantity of daily food. Physicians suggested qualitative methods very similar to those proposed in the early seventeenth century by Lessius in the *Hygiasticon*. Physician Edward Strother (1677–1737) advised that any reduced ability to labour or study after the meal was evidence of having eaten excessively.[73] Thomas Byfield warned that drowsiness or heaviness felt after a particularly heavy meal, was a sure sign of having exceeded the healthful limit. Other physical signs that eighteenth-century physicians warned readers to look out for in deciding the right quantity of food included belching, flatulence, and yawning.[74] As there were so many bodily cues for gauging the right amount of food, physicians, including those who had undertaken weighing experiments such as Keill and De Gorter, argued that it was unnecessary to make a lot of work for oneself by measuring *ingesta* and *excreta* when rules of diet could equally be determined 'by nature'.[75]

Cheyne's letters to his friend and patient Samuel Richardson offer us some further insight into the roles of both qualitative and quantitative methods for managing food intake. As Cheyne kept repeating in his letters to Richardson, he strongly believed that in matters of diet, 'the Quantity is of more Consequence than the Quality'.[76] Richardson's embodied experiences were still expected to guide him in finding the quantity which was right for him. Cheyne told Richardson he should eat as little as 'Appetite and Hunger will suffer a Man to be easy under', and Richardson was advised to eat before he felt himself growing 'low and vapourish'.[77] Additionally, Richardson was warned to 'Never fill the Stomach nor load the Nerves'.[78] While Richardson was thus feeling his way to a suitable

[73] Edward Strother, *An essay on sickness and health; wherein are contain'd all necessary cautions and directions* (London: Printed for Charles Rivington, 1725), 75, Eighteenth Century Collections Online.; T. Byfield, *Directions tending to health and long life &c* (London: Printed for the Author, 1717), 5, Eighteenth Century Collections Online.

[74] Byfield, *Directions tending to health and long life,* 5.

[75] De Gorter, cited in *Maandelyske Uitreksels,* 329–330.

[76] George Cheyne to Samuel Richardson, Dec. 23, 1741, *The Letters of Doctor George Cheyne to Samuel Richardson (1733–1743), ed.* Charles F. Mullett (Columbia: University of Missouri, 1943), 77.

[77] Cheyne to Richardson, March 27, 1739, 49 and June 20, 1739, 54.

[78] Cheyne to Richardson, April 26, 1742, 92.

quantity of food, Cheyne did not shy away from quantifying daily allowances either. In trying to convince a reluctant Richardson to reduce his *ingesta*, Cheyne referred to a former patient, a Major who was so unhealthy and unfit that the Major was carried around in a sedan wherever he went.[79] Cheyne had put the Major on a diet of water, three ounces of bread and three pints of milk a day and now the Major 'walks the Streets like a young Man', Cheyne assured Richardson. Perhaps hoping that this example would assure Richardson of the benefits of restricting his dietary intake, Cheyne prescribed Richardson a similar, though significantly more generous diet, of about 3 pints of milk and about 7 or 8 ounces of bread in a 24-hour period.[80] Cheyne's correspondence contains no reference to body weighing, which suggests that Richardson was expected to weigh his food rather than himself.

While Cheyne's quantitative approach is reminiscent of Santorio's experiment, eighteenth-century methods for quantifying daily aliment diverged from Santorio's significantly. Physicians almost unanimously advised that a weighing experiment such as Santorio's should not even be considered as a method for managing dietary intake. Most advice for determining the right dietary intake continued to refer to appetite, or feelings of fullness, heaviness, and discomfort. If the correct amount of food was quantified, this was usually done by the physician, based on his own experiment (such as Keill) or on another physician's' experiment (as Quincy did, by advising that the amount determined by Keill was right for 'most people'). Others, such as Lessius and Cheyne, landed on a suitable amount of food after trial and error, adjusting the quantity of food taken in in accordance with the body's response to that amount. Precision did not play an important role in eighteenth-century attempts to manage dietary intake, and this is especially where eighteenth-century physicians departed from Santorio. Cheyne's correspondence with Richardson is a clear example of this. While Cheyne was more willing than most physicians to quantify moderation in diet, nevertheless his advice to Richardson shows us how physicians of this period used a combination of qualitative and quantitative methods to assist their patient in reaching an amount of daily food suitable to their patient's circumstances.

While body weighing experiments practices akin to Santorio's enjoyed some popularity in late seventeenth- and early eighteenth-century medical

[79] Cheyne to Richardson, Feb. 2, 1742, 84.

[80] Cheyne to Richardson, Jan 14, 1742, 119–120.

circles, it is clear that physicians did not intend for weighing experiments to be adopted by patients. Body weighing experiments and the ensuing calculations and tables served to underline, rather than challenge, existing aphorisms concerning diet and the other non-naturals and perhaps were provided to add credibility to the existing rules, rather than to popularise new methods of health management. Described as a most troublesome and labour-intensive experiment, warnings by Quincy, Keill, and De Gorter may in fact explain why we have not found much evidence of weighing practices among the non-medical public, despite the popularity of the publications produced from these experiments. And indeed, as is evident from the correspondence between Cheyne and Richardson, there was significant concern for managing dietary quantities and physicians argued that it was important for people to have some method for measuring how much a person consumed daily lest their appetite push them beyond what was a healthful amount. It is possible that the discrepancy between the apparent interest in managing food intake and the limited evidence of weighing practices was due to the limited availability of mechanisms which could accurately weigh a human body.

Whether due to the limited availability of weighing machines or physicians' denoting it cumbersome, body weighing was not usually practised before 1760, and so people instead relied on alternative methods of determining their food intake.[81] Early eighteenth-century attempts to measure dietary quantities were not a straightforward continuation of the Santorian practice of weighing the body but instead seem to have been the result of a medical literature rife with contested interpretations of both old and new approaches to managing food intake. Although there were some significant and popular attempts at solving the issues of dietary quantities through quantifiable measurements, there does not appear to have been a clearly distinguishable shift from qualitative to quantitative approaches to managing dietary intake.

Quantification and Standardisation: General Rules of Diet?

As we have seen above, a few physicians did quantify what they argued would be a suitable amount of food for 'some' or even 'most' people. In this section, I will gauge the extent to which quantification in medicine led

[81] For an overview of the known cases of weighing before and after 1760, see Chap. 5 of this book.

to standardisation of dietary advice in the eighteenth century. The weighing experiments undertaken by Santorio as well as the eighteenth-century physicians have been interpreted as an attempt to fashion their bodies as the very basis of general rules of health.[82] However, while these physicians sought to demonstrate certain physiological rules about the body, this does not seem to have been part of an attempt to standardise dietary advice for all. The very fact that later physicians thought it necessary to re-trial Santorio's experiment to account for temperature differences suggests that the environmental context and circumstances of the individual were still believed to have a significant effect on how much a body excreted and, subsequently, how much a body needed to consume daily to make up for those excretions.

That physicians believed dietary quantities could not be standardised was evident in the manner in which the weighing experiment was recorded and explained. Eighteenth-century physicians who partook in weighing experiments, made sure to acknowledge those factors which made their bodies unique and would affect their measurements such as age, constitution, diseases, and living circumstances. Keill, de Gorter, and Robinson made readers aware of these factors by explicitly stating information regarding their age, constitution, and any ailments which they knew might affect the measurements taken during the experiment. This was important information which could help the reader decide whether he or she would excrete similarly, or differently, and whether they would therefore need to eat more, or less, or the same amount. Both the authors and readers of these reports therefore were of the understanding that the outcomes of the experiments were not reflective of all bodies. Such experiments were useful in their ability to demonstrate how varying factors, both internal and external to the body, could affect the digestive and excretory processes of bodies in a myriad of circumstances.

There is no indication that Santorio intended to standardise dietary advice by measuring how much a body excreted daily. In *Medicina statica*, Santorio does not express that his intention is to find the right measure for the 'average' person or even for 'most' people. From the writings of both Santorio and Santorio's pupil Giovanni Martino Bonomo (c.1590–1630),

[82] See Lucia Dacome especially, "Balancing Act", *Studies in History and Philosophy of Biological and Biomedical Sciences*, 43, (2012) 379–391; "Living with the chair: private excreta, collective health and medical authority in the eighteenth century," *History of Science*, 39 (2001), 467–500.

Fabrizio Bigotti deduces that Santorio's experiments likely involved men of a similar age range as those who partook in French physician Denis Dodart's (1634–1707) later experiments, who were between 25 and 45 years old.[83] As Santorio's sample constituted a relatively narrow demographic, his experiments should be seen as attempts to measure the *excreta* (and by extent, *ingesta*) of that particular group, rather than for people in general. In fact, Santorio expressly stated that the quantity excreted (through insensible perspiration in particular) varied from person to person, due to differences in constitution, age differences, diet, and environment.[84]

It was for exactly that reason that later physicians such as James Keill and Bryan Robinson thought it necessary that the same experiment was conducted again, in different environments.[85] Keill's tables of measurements demonstrated the difference in insensible perspiration between the Italians and the English which were evidence for Keill that the measurements of Santorio's experiment were not representative of an English readership. Keill re-tried the weighing experiment in an English climate so that English readers would have a better idea of how much their bodies excreted daily and therefore how much food they would need to consume to make up for this loss. Keill's finding that the body expended 'more than one and thirty Ounces' by perspiration fell short of Santorio's 50 ounces.[86] Which, Quincy commented, was hardly surprising, 'for where [Santorio] made his Trials was vastly warmer than where Dr. Keil [sic] made his, and therefore would a great more, for that very reason, be wasted by the Surface of the Body in insensible Steam'.[87] It was therefore understood that not all of Santorio's aphorisms were true for all people, and there is no indication that people had assumed they were.

Though differences in temperature seems to have been the primary grounds on which Keill justified a re-trial of the experiment, there are

[83] Fabrizio Bigotti, "Intensity Meters: New Notes and Discoveries on the Invention of Early Modern Precision Instruments", *Early Science and Medicine*, 28 (2023), p.614; Barry and Bigotti, *Santorio Santori and the Emergence of Quantified Medicine*, 4, and 17–18.

[84] Santorio, *Medicina Statica* (1720 version), 47–48.

[85] James Keill, *Essays on the several parts of the animal oeconomy* (London: Printed for George Strahan, 1738), unpaginated preface, Eighteenth Century Collections Online.

[86] See Ibid., 242, versus Santorio, *Medicina statica* (version 1737), 50.

[87] Quincy, in *Medicina Statica: being the aphorisms of Sanctorius … To which is added Dr. Keill's Medicina Statica Britannica, with comparative remarks … by John Quincy.* 2nd edition (London: Printed for W. and J. Newton, 1720), 323. Eighteenth Century Collections Online.

indications that he considered other factors also. In his preface, Keill included a description of himself which featured the same factors frequently mentioned by physicians in this period when considering how much a person ought to eat, exercise, or otherwise manage their non-naturals, namely age, diseases, and constitution. Keill specified that the bulk of his experiment was undertaken while he was between the ages of 30 and 40, which placed him in a particular age group believed to be physiologically different from children and people of old age. With this in mind, we ought to consider that Keill might have mentioned his age here to make clear that some of his findings would not apply to readers of other ages. Keill does not state explicitly that his finding that the body perspires 31 ounces does not apply to bodies of different ages than his, but he does say in the aphorism immediately following that perspiration levels are varied according to age, which strongly implies that this is the message he intended to convey to his readers.[88] Three decades later, Irish physician Bryan Robinson similarly prefaced his treatise by mentioning a few 'particulars' about himself. He summarised: 'I am now, in May 1747, in the 68th Year of my Age. The Length of my Body is 63 Inches: I am of a sanguine but not robust Constitution, and at present I am neither lean nor fat'.[89] Robinson also included information regarding a particular disorder he had which had affected his measurements.[90] The fact that both men mentioned similar 'particulars' even though their experiments were conducted many years apart suggests that there was little change in which factors were thought to affect *excreta* and the significance ascribed to them in relation to the undertaking of such an experiment.

Keill similarly informed the reader that he had suffered from a Nephritic complaint, which, he hastened to add, 'never retarded the Urine'.[91] Had Keill's Nephritic complaint affected his urine, his findings for the proportions between the different *excreta* would have deviated from his usual readings as well as from the readings of any other person unaffected with urinary issues. Finally, Keill specifies that his constitution is 'moist and warm' and admits that 'Perhaps there may be other proportions between the Urine and Perspiration in other Constitutions'.[92] Clearly, Keill was

[88] Keill, *Medicina Statica* (version 1737), 323

[89] Bryan Robinson, *A dissertation on the food and discharges of human bodies* (Dublin: Printed by S. Powell, 1747), Eighteenth Century Collections Online.

[90] Ibid., 35.

[91] Keill, Essays on several parts of the animal oeconomy, unpaginated preface.

[92] Ibid.

aware of the possible effects of these factors on the measurements and thought it important to provide this information to help the reader understand any differences between his and Santorio's findings. As Dacome has demonstrated, criticism of these experiments often hinged on whether readers believed the experimenter had been successful in minimising variables which could affect the findings.[93] Keill's colleague Rye had suggested that Keill's irregular bedtimes and interruptions to the experiment had obstructed his ambition to 'bring this branch of practical medicine to perfection', and Dutch experimenter De Gorter 'suspected' that Keill's 'weak or sickly body' had been the cause of the discrepancies between Keill's and Santorio's measurements of daily perspiration.[94]

It is possible that the experimenters sought to present their bodies as close to what they thought was the 'average' body of their readership. There is evidence to suggest that authors of health guides believed their readership to be a fairly homogenous group, and it is likely that Keill and Rye expected the readership of their aphorisms to be similar to that of the general health guides. Keill might have thought his measurements to be representative of men of his age, living in the same climate and under similar levels of physical exertion. Keill did emphasise that 'the proportion of daily Food, suitable to a Body *in the fore-mentioned Circumstances,* is about four Pound'.[95] But, added Quincy to this aphorism, 'tho' this [quantity] settled by [Keill] is very likely to be the Quantity convenient to most People', 'what Quantities of Food will best preserve such a standard, is best within every ones own Experience'.[96] Keill himself had similarly advised his readers to consider their own situations rather than simply copy his measurements. Readers of his tables of data were expected to be their 'own Judge of the truth of the Aphorisms', and 'if these do not answer his expectation, it may not be disagreeable to him to look over the Tables, and by comparing one day with another, to pick out what he pleases'.[97] In other words, readers of these experiments were expected to take an active role in the constructing of their own dietary rules, for which the tables served only to demonstrate the effects of factors such as dietary

[93] Dacome "Living with the chair", 484.

[94] Ibid.; De Gorter, cited in *Maandelyske Uitreksels,* 325–326.

[95] Keill, *Medicina Statica* (version 1720), 336.

[96] Quincy, *Medicina Statica* (version 1720), 336.

[97] Keill, *Essays on the several parts of the animal economy*, unpaginated preface.

choice, exercise levels, illnesses, and local climates on the balance between the *ingesta* and *excreta*.

The season of the weighing experiment came to a close with the final one conducted by William Stark in 1769, the findings of which were published posthumously in 1788. Stark's motivations were slightly different from those of the former experimenters, which perhaps reflects a more general decline in interest in this minute and tedious method of health monitoring, something which has already been noted by Dacome.[98] Foregoing the traditional focus on insensible perspiration, Stark sought to demonstrate the effects of various aliments on the human body, ultimately to prove that a 'pleasant and varied' diet was as conducive to health as the spare diet promoted by Cornaro.[99] Moreover, Stark wondered openly whether the exactness of the discharges of the body 'were more a matter of curiosity than of use'.[100] Dismissing the quantitative methods to which his predecessors ascribed so much authority, Stark directed his attention to observing the different effects of food on the body, measured by noting whether the body agreed or disagreed with certain aliments, which foodstuffs invigorated or indisposed the body, and which allayed or created an appetite.[101] Despite his conviction that this kind of knowledge was of more use than measurements of the discharges, Stark did include these and a myriad other measurements in his tables, in order to 'avoid the appearance of entertaining a doubt', he explained.[102] While Stark therefore still ascribed some authority to the weighing experiment, his approach to establishing rules of diet constituted a careful but deliberate departure from the early eighteenth-century focus on the exact, the minute, and the quantifiable. Similar experiments into the dietary needs of the human body would continue to thrive in the medical world, but they would do so in a very different context.

The popular dietary advice of this period echoed the arguments made in the treatises written by the experimenters in that the concept of a healthful quantity of food was relative and depended on—especially—constitution,

[98] Dacome, "Living with the chair", 488.

[99] William Stark, *The works of the late William Stark, M.D. consisting of clinical and anatomical observations, with experiments, dietetical and statical, revised and published from his original Mss by James Carmichael Smyth* (London: Printed for J. Johnson, 1788), 89, Eighteenth Century Collections Online.

[100] Ibid., 91.

[101] Ibid.

[102] Ibid.

age, and any illnesses. Maynwaring argued that 'a set quantity or measure of *meat* or *drink*, cannot be prescribed as a *general rule* and *observation* for all to follow', because people have different constitutions, infirmities, levels of strength, and physical labour, and so '[W]hat is convenient for *one*, is too much for *another*, and too little for a *third*'.[103] Physician Byfield agreed, citing similar factors: 'There cannot be one exact Measure for all people, not for the same Person, at all seasons: But must vary, according to the Diversity of Constitutions, Ages, Seasons of the Year: Disposition, or Indisposition, of the Stomach, and other Infirmities or Circumstances'.[104] Age appears to have been an especially important factor to physicians of this period as they tended to address their dietary rules especially at those who had passed a certain age threshold. Cheyne was very specific in his advice regarding quantities in the diets for older readers and recommended all persons over the age of 50 to reduce their dietary intake by a quarter every 10 years.[105] Most other physicians were not keen on exact measurements and recommended general temperance instead, but the idea that age was a primary factor in dietary decision making was a common feature. The theory that strict temperance was necessary for those who had passed their 'meridian' seemed to have gained popularity after the publication of Cornaro's treatise, as physicians continued to refer back to Cornaro's case as an example of the health benefits of a minimal diet. Physician John Hill argued (1714–1775): 'The quantity of food at a meal should be less for old persons than for young: and the older they grow, the more this should be diminished. This was the practice of Hippocrates; and by the observance of it Cornaro lived to his extreme age', he added to strengthen his point.[106] Moderation was important in all ages of life, 'but', Hill advised, 'most of all in the old man's life'.[107]

Differences of constitutions remained relevant to discussions about diet, and this applied to discussions about quantities as well as qualities. Still in the 1770s, readers of health guides found it argued that it was 'the

[103] Everard Maynwaring E, *Vita sana & longa the preservation of health and prolongation of life proposed and proved* (London, Printed by J. D.,1669), 55–56.

[104] T. Byfield, *Directions tending to health and long life &c* (London: Printed for the Author, 1717), 4–5. Eighteenth Century Collections Online.

[105] Cheyne, *An essay on regimen*, xxx.

[106] John Hill, *The old man's guide to health and longer life: with rules for diet, exercise, and physick*. 2nd edition (London: Printed for M. Cooper, 1750), 11, Eighteenth Century Collections Online.

[107] Ibid.

Duty of every Person to study his own Constitution, and to Eat and Drink of such Things, and in such Quantities, as he finds to agree best therewithin'.[108] However, the term constitution underwent some changes, with later physicians using it not only to describe certain humoural mixtures but also body shapes and sizes, and the strength or weakness of the solids of the body.[109] Language changed as medical theory shifted, and this idea, which was based in humoural theory, was increasingly explained through chemical or mechanical theories and terms. For example, Mackenzie argued that the quantity of diet should be based on what the vessels could handle, which differed per individual constitution: '[e]very body has its particular latitude, that is, its vessels may be stretched to a certain degree, and yet restore themselves. Four pounds of meat and drink is as much, or more than some constitutions can well bear, whereas others can take in eight pounds without any inconvenience'.[110] Though expressed through iatromechanical explanations of the body, in practice physicians continued the tradition of adjusting dietary advice carefully to the strengths and weaknesses of individual bodies.

The non-naturals were to be managed very differently for a sick person than for a healthy person. For healthy bodies, the primary purpose of diet was to provide exactly enough nourishment to replenish what the body had lost during the day. Physicians espousing iatromechanical ideas argued that with every single action, the fibres of the body 'by their attrition one against another, file off from one another vast numbers of little seperable parts' causing the fibres to grow weaker over time until they were unfit to perform their function.[111] The only way to slow this process was by repairing them. 'Now 'tis our Food, whether Fluid or Solid, that furnishes us with this Supply', Wainewright explained, '[h]ence we may deduce the necessity both of taking in Food, and also of the circulation of the blood, for if either of these were wanting, there would be no means left of repairing the loss the Fibres sustain by their daily contraction'.[112] Eating too

[108] Anon., *Directions and observations relative to food, exercise, and sleep* (London: Printed for S. Bladon, 1772), 4, Eighteenth Century Collections Online.

[109] Ibid.

[110] Mackenzie, *The history of health*, 275 – 276.

[111] Wainewright, *A mechanical account of the non-naturals*, 149; William Forster, *A treatise on the various kinds and qualities of foods: with aphorisms of health, or, rules to preserve the body to a good age* (Newcastle upon Tyne: Printed by John White, 1738), 3–4, Eighteenth Century Collections Online.

[112] Wainwright, *A mechanical account of the non-naturals*, 149–150.

little would lead to continual wasting of the body's fibres, until they would no longer be able to fulfil their primary function: to circulate the body's fluids. The body would then suffer a 'langour' or insufficient strength to perform the bodily functions and eventually, to support life.[113] Too large a quantity of food, on the other hand, overwhelmed the stomach and its ability to digest matter into perfect chyle, causing corrupted matter to enter the blood and cause inflammations and disruption to the digestive system. This disrupted system would again mean that the body's solids would not receive their due nourishment, leading to weak fibres, malnourishment, and inability for the body to sustain itself and carry out its functions.[114] For healthy bodies, therefore, the primary purpose of diet was to provide exactly enough nourishment to replenish what the body had lost during the day while avoiding excesses which could overwhelm the digestive system.

Those who were sick required a course of diet specifically designed to cure that illness. In traditional Galenic medicine, foods were grouped together according to their primary qualities, their degree of heat, coldness, moisture, and dryness. This meant eating food items which were of an opposite nature to the excess humour which had caused the patient to be unwell. In the seventeenth and eighteenth centuries, however, the focus became more on the relative ease of digestion of foodstuffs.[115] Diets prescribed by seventeenth- and eighteenth-century physicians to people who were suffering from illness or who were considered to have 'weak digestion' due to old age or a weak constitution increasingly focussed on food which was easily digested, though many physicians did still pay attention to the humoural qualities of foodstuffs to at least some degree. Cancer, for example, was a disease widely believed to have its origin in humoural dysfunction, specifically the 'burning or stagnation' of black bile (also referred to as melancholy), one of the four bodily humours.[116] As Alanna Skuse has shown, early modern physicians therefore prescribed diets specifically aimed at reducing excess black bile and quelling

[113] Francis De Valagin, *A treatise on diet, or, the management of human life; by physicians called the six non-naturals* (London: Printed for the author by J. and W. Oliver, 1768), 81–84, Eighteenth Century Collections Online.

[114] De Valagin, *A treatise on diet* 168.

[115] See Chap. 2 of this book, as well as: Michael Stolberg, *Experiencing Illness and the Sick Body in Early Modern Europe* (Basingstoke: Palgrave Macmillan, 2011), 122–123.

[116] Alanna Skuse, *Constructions of Cancer in Early Modern England*, Palgrave Studies in Literature, Science, and Medicine (Basingstoke: Palgrave Macmillan, 2015), 98–99.

inflammation in the bodies of those suffering from cancer. These diets focussed on avoiding foodstuffs which were described as strong flavours and thick consistency and instead consuming foods which had 'cooling' and 'moistening' properties, both to combat the 'dry' humour and to 'cool' inflammation.[117]

There are some indications however that the eighteenth century witnessed somewhat of a shift towards diets designed to be suitable for many and not just the few suffering from particular diseases. This seems to be the case especially for diets which were 'cooling' and 'moist', which was generally thought to be of benefit to all who suffered from some form of 'weakened digestion'. It appears that eighteenth-century authors of dietary advice literature argued increasingly focussed on curing the symptoms which people shared, rather than the ailments which set them and their dietary needs apart. For those people who suffered from weak digestion, a general diet of 'convalescent food' was warranted. Ken Albala has shown that convalescent cookery (the ingredients, culinary techniques, and dishes thought to be appropriate for a category of people who were weak, recovering from illness, or thought to have impaired digestive systems) remained constant throughout the early modern period, remaining largely unaffected by shifts in medical theory.[118] Convalescent cookery was primarily focussed on the idea that people whose digestive abilities were affected by illness, old age, or a weak constitution, needed foods which were easily digested.[119] As we have seen in the previous chapters, these foods which were 'easily digestible' were often described as being 'moist', 'cool', or 'light', and even of these foods, patients were advised to eat as little as they were comfortable under. This is evident in the works of Cornaro and Cheyne, whose diets both resembled what Albala described as convalescent cookery. Cornaro described himself as having been 'addicted' to 'Intemperance', which, in combination with his generally 'ill Constitution', caused him many diseases, including stomach pains, fevers, and the gout.[120] Cornaro was advised by his physicians to live 'a sober and orderly Life', and under their direction, Cornaro switched to a diet such as belonged to a sick person, eating and drinking only what was

[117] Ibid., 99.

[118] Ken Albala, 'Food for healing: Convalescent cookery in the early modern era', *Studies in History* and *Philosophy of Biological and Biomedical Sciences*, Volume 42, Issue 2 (June 2012), 323–328.

[119] Ibid., 323.

[120] Cornaro, 'of the Benefits of a Sober Life', annexed to *Hygiasticon*, 8.

'wholesome' and only very small quantities.[121] Cheyne argued that the very source of 'chronical distempers' was 'Viscidity in the Juices' and 'Over largeness of their constituent Parts', both of which were begotten by 'excess in Quantity' and 'ill condition of our Meat and Drink'. According to Cheyne, the cure for most, if not all chronic distempers was therefore a diet consisting of 'light' and 'vegetable' food, restricted to 8 ounces of meat, 12 ounces of bread or 'vegetable food', and about a pint of wine daily.[122] As we have seen in the previous chapter, Cheyne advised that such a diet of the 'least' amount of the food and the 'lightest' kind of food could 'supply the place of any specific regimen of diet', which suggest that certainly by the eighteenth century, dietary advice, both in quality and in quantity, became somewhat homogenised for all those who were ill, recovering from illness, old, or generally 'weak' or of 'weak digestion'.

The circumstances of one's life, especially those relating to occupation and exercise levels, were also believed to affect digestive abilities and perspiration levels, and therefore constituted another important factor in deciding how much food a person needed. Much like today, physicians argued that those who exercised much, through their occupations or otherwise, needed more food than those whose lifestyles were mostly sedentary.[123] 'If you eat much, you must work, or exercise much; and if you use much Exercise, you must eat heartily', Strother advised, because these 'necessarily infer one another; Exercise wastes us, and Victuals repair us'.[124] Learned men, Italian physician Bernardino Ramazinni (1633–1714) argued in his *Treatise of the Diseases of Tradesmen*, were 'as Slothful and Idle in their Body as they are Active and Busie in their Mind and Brain' and almost all of them suffered the inconveniences of a sedentary life.[125] These consequences included a 'weakness of the Stomach', of which, Ramazinni found, all the men of learning complained.[126] 'There's no hard

[121] Ibid., 9–12.

[122] Cheyne, *An essay of health and long* life, 19–20 and 33–34.

[123] See, for example, Byfield, who argued that those 'who work like horses, should feed like horses', while those 'employed in study' ought to eat less, in *Directions tending to health and long life*, 11.

[124] Edward Strother, *The family companion for health, or, plain, easy and certain rules* (London: Printed for F. Fayram, 1719), 46, Eighteenth Century Collections Online.

[125] Bernardino Ramazzini, *A treatise of the diseases of tradesmen, shewing the various influence of particular trades upon the state of health* (London: Printed for Andrew Bell et al., 1705), 246–247, Eighteenth Century Collections Online.

[126] Ibid., 248.

Student almost but what complains of his Stomach', he found, 'For while the Brain is imploy'd in digesting … the Stomach can't but make an imperfect Digestion of the Aliment'.[127] A sedentary occupation was not limited to the elite of course, as shoemakers and tailors also spent much of their days being seated. Though their bodies suffered from bad posture, yet people in these trades used their hands and were not without exercise. While Ramazinni's sedentary group thus included people from all levels of wealth, many health guides were generally aimed at the more elite echelons of society. Owing to a combination of a sedentary profession and a leisurely lifestyle, many an elite person was believed to be suffering from an impaired digestion while also lacking the level of perspiration a more active person would enjoy, both of which were factored into the dietary advice found in the guides. The advice given in these guides therefore did not apply to the population in general but was carefully crafted for those whom physicians believed needed very carefully quantified diets. For example, Byfield explicitly warned that his recommendations of water-gruel and periods of abstinence constituted a 'Schollar's Allowance' which was not suited for all people.[128]

Cheyne focussed especially on age and physical activity when discussing dietary quantities while factors such as constitution and sex received far less attention from him. In fact, Cheyne seems to have implied that past the age of 50, sex difference ceased to be a factor worth considering at all. He argued that 'the great critical Alteration' that happens 'in or about Fifty', was often wrongly believed to be particular to women only.[129] The consequences of the weakening of the fibres and the digestion made this age the 'True Time and Season of Life' for paying attention to diet, for both sexes alike.[130] Elizabeth Williams noted that 'the equivalence between the male and female "economy" became more pronounced as the years passed', and dietary advice put more emphasis on what the patient needed based on age, more so than on gender—though this view became more prevalent in the nineteenth century, she argues.[131] It appears however that

[127] Ibid.

[128] Byfield, *Directions tending to health and long life*, 11.

[129] Ibid., xxxii.

[130] Ibid., xxxiii.

[131] Elizabeth Williams, "Eating after the climacteric: food, gender and ageing in the long eighteenth century", in *Lifestyle and Medicine in the Enlightenment,* edited by Kennaway, James and Knoeff, Rina. London: Routledge (2020), 78, https://doi.org/10.4324/9780429465642.

at least some eighteenth-century physicians already believed there to be enough equivalence between male and female elderly patients that they could be prescribed similar dietary rules. Most physicians seemed to have made little distinction of gender in their dietary advice for old people, although we must keep in mind that it was customary for physicians to refer primarily to the male body in their discussions about the body. The reduced ability of old people to digest their food in a timely manner was a factor which physicians seemed to believe was shared between many, if not most old people, and justified a certain degree of generalisation in dietary advice aimed at all people who were old, regardless of their gender, individual constitutions, or circumstances of life. Hill, for example, argued that his 'general rules' would 'hold universally, because they are calculated for old people, merely as they are old; not as they are of one another constitution'.[132] This suggests that as far as Hill was concerned, the physiological issues shared between old bodies overrode any other non-natural or physiological differences when it came to dispensing dietary advice.

Why did physicians such as Lessius and Cheyne attempt to quantify a dietary daily allowance, if such a quantity could not possibly apply to all or even most people? Both authors were attempting to quantify the right amount of daily food for a group of people they argued shared much of their dietary needs. Lessius's recommendation for 'twelve or fourteen Ounces of Food a-day' was specifically meant for those 'who are much advanced in Years, and for those also, who are of weak Complexions'.[133] Cheyne's *Essay on Health and Long Life* specifically addressed 'the Sickly and the Aged, the Studious and the Sedentary, Persons of weak Nerves, and the Gentlemen of the learned Professions'.[134] His *Essay on Regimen* was written to relieve 'the Scorbutic, Gouty, Consumptive Valetudinarian-low-livers' and those 'whose vocation and occupation is sedentary and studious'.[135] People belonging to these groups, namely the old, the sedentary, and the generally 'sickly', were argued to possess weak digestive abilities, and it was in dietary advice aimed at these groups in particular that we find attempts to quantify a healthful daily dietary intake. Neither Lessius nor Cheyne sought to quantify the dietary needs of the 'average' human body. Though the works of Lessius and Cheyne constitute attempts to

[132] Hill, *The old man's guide to health and longer life*, 10.
[133] Lessius, *Hygiasticon*, 22.
[134] Cheyne, *An essay of Health and Long Life*, xvi.
[135] Cheyne, *An Essay on regimen*, xxx.

quantify and, to a lesser extent, standardise the diet of people grouped together based on their shared metabolic characteristics, it is important to point out that generally, physicians of this period echoed the classical advice that dietary quantities had to be determined on a case-by-case basis, accounting for a wide range of individual factors including age, constitution, environment, and even personal dietary preferences.

Seventeenth- and eighteenth-century health guides continued to provide dietary advice which operated on the assumption that there could be no health rules which applied widely. Physicians echoed traditional axioms which warned that how much a person should eat depended on many factors including age, constitution, season, occupations, and illnesses. In fact, any physician who wrote rules which seemed a little too general could count on his colleagues' criticisms. Readers of these guides would have found it stated explicitly that they were expected to formulate their own regimen from the myriad of different sections on diets for choleric or sanguine constitutions, for acute and chronic illnesses, for old and young bodies, for warm and cold climates. In such a medical climate, recommending certain quantities of food was generally believed to be neither practical nor desirable. That being said, we can clearly discern a trend in the late seventeenth and early eighteenth century where physicians sought to quantify the dietary needs of their own bodies and of the bodies of people with generally weak digestive abilities. We can also discern a shift towards the standardisation of dietary rules of people who were believed to require a diet consisting of less food than the 'healthy' individual, such as people of old age, of sedentary lifestyles, and of weak constitution or with weak digestive abilities.

Conclusion

This chapter began with two questions. First, I asked whether the period under consideration witness a shift in dietary advice from using appetite as a guide to using a measure for determining how much a person should eat daily. As we have seen, the influx of new foodstuffs and French cooking trends had physicians concerned that people could no longer discern a healthy from a spoiled appetite. Delicacies and recipes from all over the globe enticed eaters to take in more than they strictly needed, ignoring their bodily cues of fullness. Hence, physicians called for rules to guide readers of their health guides. These developments, in combination with the popular school of thought known as iatromechanism, led to an increased interest from physicians in quantifying how much daily food the human body needed. The weighing experiment undertaken by Santorio in the late sixteenth century enjoyed renewed interest from seventeenth- and eighteenth-century physicians looking at past authorities for solutions to contemporary problems. Santorio's method for quantifying how much weight the body lost due to daily excretions and how much the body needed to ingest to make up for that loss, became much discussed and even re-tried over the course of the late seventeenth and eighteenth centuries. But while quantification of the body's metabolism was lauded by many as a sure and precise method for understanding the body's digestive processes, physicians also acknowledged that a measuring experiment was difficult to execute and ultimately impractical as a method of determining how much or how little a person should eat daily. Though quantification as a topic was popular in late seventeenth- and early eighteenth-century discussion about diet, in practice, food measuring did not replace other qualitative methods for determining dietary intake. Authors of health guides continued to advise their readers to continue to listen to bodily cues of fullness, often in combination with quantitative methods, such as food measuring or adhering to advice which was based on a measuring experiment. There was no significant shift from using qualitative to quantitative methods for managing diet.

My second question was whether quantification in dietary advice led to standardisation of dietary advice. As I have shown, however, physicians continued to repeat that no quantity or measure could possibly apply to all or even most people. How much a person needed, they argued, depended on individual factors including the constitution, age, occupation, ailments, and environmental circumstances of the patient, and health advice continued to be organised in a manner which expected readers to choose from the myriad of rules the ones which suited their bodies and lives best. The eighteenth-century weighing experiments therefore should not be taken as evidence of physicians' attempts to standardise rules of health. Although Santorio's experiment had the medical world quite convinced of the value of static medicine, his aphorisms were not adopted thoughtlessly. In fact, physicians quickly decided that re-trials were necessary to accommodate the effects different temperatures had on the body. A man living in Italy would perspire differently from a man living in England, and so these men would require different quantities of food. Although the experimenters' focus was predominantly on environmental differences, they too explicitly warned that measurements of *ingesta* and *excreta* differed per constitution, age, occupation, and ailments, as well as local temperatures. Similarly to the health guides, readers of these tables of measurements were encouraged to read critically and with their personal circumstances in mind. Readers were informed of the constitution, age, ailments, and living circumstances of the experimenter and would therefore be able to judge how close or different their measurements would be to the experimenter's. While the experimenters sought to prove how bodies worked in terms of their *ingesta* and *excreta*, they did not lead to any standardisation in health advice. The tables were, after all, evidence of how the usual factors could affect the *excreta—and* therefore the *ingesta*. Bodies functioned, physiologically, the same, yet they were still considered very much part and parcel of their unique environments and affected by their living circumstances.

Nevertheless, we do find evidence of some physicians, notably Cheyne, prescribing dietary rules which he argued would apply to large groups of people, if not most of his readers. Physicians such as Cheyne seemed confident that certain dietary rules, especially relating to the quantity of food, applied to most people who had weak digestive abilities, such as old people, people with sedentary lifestyles, and people who could generally be described as 'sickly'. It appears that over the course of the eighteenth century, physicians increasingly prescribed and treated based on the disease of the patient, rather than on the patient's humoural imbalances. Shared symptoms, such as having difficulty digesting, could therefore lead to similar dietary treatments. Some health guides were written with a specific readership in mind, and so it is possible that physicians believed that the bodies of their readers were likely to be quite similar, at least in their ailments, their age, their occupation and environmental circumstances. This allowed physicians to be quite general in some aspects of their health advice—but it is important to note that these general rules were accompanied by warnings that these recommendations were not a golden standard to be adopted by all. In general, physicians remained clear in their conviction that general rules were impossible and undesirable.

The way in which quantification affected seventeenth- and eighteenth-century dietary advice was unique to the period under consideration. The popularity of iatromechanical thought, with its focus on mathematical rules for health and the body, had to contend with the ongoing popularity of Galenic ideas, as well as the difficulty popularising experiments, which required a combination of time, effort, and a carefully controlled environment. In addition, the technology to measure the body (the 'Sanctorian' weighing chair, as we will discuss in Chap. 6) was not available to most people, which restricted this method for measuring *excreta* and *ingesta* to a few physicians and nobles able to procure such a device. As we shall see in the last chapter of this book, the popularity of Santorian ideas had started to wane by the time personal weighing devices became more widely available. Ideas about what constituted an 'average' body and how much food an average body needed to consume would continue to be discussed in the late eighteenth and especially early nineteenth centuries, but these discussions would take place in a very different context and for different purposes. Despite some very strong studies into this very area of study, more research is still needed to establish how different medical, political, and social developments interacted to create a new medical culture which prioritised general over individual dietary rules. The seventeenth- and eighteenth-century quantification theories and experiments seem not to have naturally evolved into nineteenth-century efforts to quantify human dietary needs. The experiments undertaken by Keill, Robinson, and Stark were undertaken in a unique historical milieu and for very different motivations than the experiments in nineteenth-century workhouses, prisons, and hospitals. It would take a range of circumstances and developments, inside and outside of medicine, to change this.

CHAPTER 4

'The Debauchery of Inaction': Idleness and Sedentarism in Health Advice

The field of medical history has thus far favoured dietetics over exercise when it comes to researching health concerns of the past, though some excellent studies have brought our attention to the central place exercise took in early modern health literature.[1] These studies have highlighted the complexities of how and why certain activities were or were not considered exercise, as well as the many different medical, social, and cultural developments and factors which caused shifts in which activities were recommended as exercise. In *Healthy Living in Renaissance Italy*, Sandra Cavallo and Tessa Storey found that by 1600, definitions of what counted as healthful exercise gradually shifted away from what was still recommended in late medieval health literature and that these shifts were closely connected to new views of the ideal male body and patterns of genteel life

[1] A more extensive overview of the historiography of sport and exercise history can be found on pages 26–30 of this book, but some particularly good studies include Sandra Cavallo and Tessa Storey, *Healthy Living in Renaissance Italy* (Oxford, Oxford University Press), 145–177; Rebekka von Mallincrodt and Angela Schattner (eds.) *Sports and Physical Exercise in Early Modern Culture: New Perspectives on the History of Sports and Motion* (Abingdon, Routledge, 2016); James Kennaway and Rina Knoeff, '"For it is the debilitating fibres that exercise restores": Movement, morality and moderation in eighteenth-century medical advice literature' in James Kennaway and Rina Knoeff (eds) *Lifestyle and Medicine in the Enlightenment: The six non-naturals in the long eighteenth century* (Abingdon, Routledge, 2020), 108–111.

M. Wübbels, *Quantification and Measurement in British Health Advice, c. 1670–1830*, Palgrave Studies in Medieval and Early Modern Medicine,
https://doi.org/10.1007/978-3-032-32760-4_4

brought about by an emerging aristocratic culture.[2] Cavallo and Storey found that in Italy, the early development of court life and the demilitarisation of the nobility and of urban society at large had a significant effect on ideas of what constituted healthy forms of exercise. Though these developments took place in a different country, and some time before the period under consideration here, Cavallo's and Storey's findings underline the potential effects of social and cultural changes on medical advice literature, which is also apparent in eighteenth-century British medical advice literature, as we will discuss below.

Studies of eighteenth-century exercise advice have similarly found considerable shifts in ideas of what exercise was and how it should be practised. James Kennaway and Rina Knoeff discovered that the focus on gentle and moderate exercises which had been so prevalent in sixteenth-century Italy (and, as we will discuss, in early eighteenth-century Britain) gradually started to give way in the late eighteenth and early nineteenth century to more vigorous exercises designed to strengthen the body.[3] Kennaway and Knoeff attribute this late eighteenth-century shift to a combination of new medical theories and concerns about changing lifestyles. As we shall see, these two factors also played a role in the shifts in early eighteenth-century exercise advice. Despite this excellent recent work, historians have not yet given exercise the place it deserves in any analysis of eighteenth-century health cultures. This disparity between interest in diets and exercise may be due to developments in fields relating to the history of the body and especially on issues to do with fat bodies. With concerns around fatness and dieting ever increasing amidst our modern 'obesity epidemic', the history of diets has been given an impetus that the history of exercise has not yet enjoyed, despite the two practices of diet and exercise being so intrinsically linked in our modern health culture. We still have a very limited idea of how exactly eighteenth-century people exercised, for what reasons they exercised, and how medical theories influenced exercise advice. This chapter seeks to redress that imbalance by assessing exercise advice and practices discussed in the many health guides which were published throughout this period and comparing this to advice about diet and body weighing, in order to analyse how these topics were intricately connected.

[2] Cavallo and Storey, *Healthy Living in Renaissance Italy*, 145–177.

[3] Kennaway and Knoeff, "For it is the debilitating fibres that exercise restores", 108–111.

Throughout the early modern period, exercise was generally thought of as beneficial to the health of the body. It provided a natural and safe way for the body to excrete the matter taken in by eating and drinking, and for that reason had always been part of that group of six non-naturals on which health relied. As this chapter will demonstrate, the explanations for why exactly exercise was so beneficial shifted somewhat as medical theories changed, but regardless of school of thought, physicians usually mentioned its advantageous effects on the digestive processes, from proper 'concoction' in the stomach to timely evacuation of the bowels, all of which helped prevent congestions in the body.[4] The importance which was ascribed to exercise, however, did change, the reason for which, I will argue, can be found in shifts in medical theory taking place between the late seventeenth and early eighteenth centuries.

In the late seventeenth century, a medical school of thought known as iatromechanism became particularly popular.[5] It was in this period that anatomists observed through their microscopes that the body was composed of minuscule interconnected threads, or 'fibres', which physicians of the iatromechanical school of thought argued to perform a vital role in the functioning of the body's economy.[6] These fibres were understood to be the very threads that muscles and veins consisted of, and to keep the muscles and veins in working order, these fibres needed to remain in a healthful state. It is around the same time that physicians started to view the body as wholly composed of vessels through which the body's various fluids flowed. Indeed, Bath-based physician George Cheyne described the body as 'a compages or contextures of pipes, an hydraulic machin, filled with a liquor'.[7] Various theories and discoveries provided the impetus for to this focus on the vessels of the body, including William Harvey's (1578–1657) discovery of the circulation of the blood, René Descartes's (1596–1650) conception of the

[4] Everard Maynwaring, *The method and means of enjoying health, vigour and long life adapting peculiar courses for different constitutions, ages, abilities, valetudinary states, individual proprieties, habituated customs, and passions of mind* (London: Printed by J.M. for Dorman Newman, 1683),138–139, Early English Books Online.

[5] A helpful overview of the shifts and developments of medical theories and schools in the early modern period, including the development of the iatromechanical school of thought, can be found in Ruben E. Verwaal, *Bodily Fluids, Chemistry and Medicine in the Eighteenth-Century Boerhaave School* (London: Palgrave Macmillan, 2020), 3–4.

[6] For a comprehensive overview of fibre and its prominent role in eighteenth-century understanding of the body, see Hisao Ishizuka, *Fiber, Medicine, and Culture in the British Enlightenment* (New York, Palgrave Macmillan, 2016).

[7] Cheyne, Essay on regimen, p. ii.iv.

body as a hydraulic machine, and Marcello Malpighi's (1628–1694) discovery of the capillary blood vessels in the lungs, all of which underlined the importance of these vessels to the functioning of the body.[8] All of these vessels consisted of these interwoven fibres, which resulted in fibres taking centre stage in late seventeenth- and eighteenth-century health advice.

Essential Exercises

This new privileged position that the vessels of the body enjoyed was reflected in an increased emphasis on exercise advice, with Francis Fuller's (1670–1706) *Medicina Gymnastica* being perhaps the first popular health guide to discuss exercise primarily. Fuller explicitly prioritised the role of the vessels over the fluids in the maintenance of health.[9] In this treatise on exercise, Fuller lamented that, in the past, too much of health was attributed to the fluids and too little to the solids.[10] The vessels of the body were not considered a passive part of the body but were ascribed an active role in the circulation of the body's fluids. The fibres that the vessels consisted of were understood to possess an elastic quality. The pressure applied by the fluids on the elastic fibres resulted in the fluids being forced or 'propelled' forth through the vessels. If the fibres lost their elastic quality, they would lack the force to drive the fluids forth, which would result in a slowed circulation of the fluids, which was detrimental to health.[11] The more powerful the assistance from the elastic fibres was, the swifter the blood circulated through the body. Though 'elasticity' was generally used to describe this force, 'tone' and 'firmness' were also used to indicate the fibrous power which propelled the blood forward.[12] '[I]n Elasticity alone, the force, power and pleasure of life and the animal functions, consists', Cheyne insisted, thus elevating the fibres to a level of medical relevance they had not enjoyed before.[13] This did not mean that the fluids were no

[8] Ibid., especially 7–9.

[9] Francis Fuller, *Medicina Gymnastica: or, a treatise concerning the power of exercise* (London: Printed by John Matthews, 1705), Eighteenth Century Collections Online.

[10] Ibid., preface.

[11] Ibid., 13.

[12] Which terms were preferred depended on period and different schools of thought, and although they were not interchangeable, these terms all referred to such a force, albeit with different aspects to it.

[13] George Cheyne, *An essay on regimen* (London: Printed for C. Rivington, 1740), xii, Eighteenth Century Collections Online.

longer considered important to health but rather that '[T]hey have a necessary dependence upon each other', physician Bernard Lynch explained:

> For as the distention and power of restitution in the vessels, owing to them being duly moisten'd by some convenient fluid separated and dispensed from the blood ... so likewise that due constitution of the blood ... depends upon its certain degrees of fluidity, which are always as the contraction of its circulating vessels.[14]

But, Lynch added, 'even in this the solids' have the principal share, because ... that constitution enabling it to afford such a power, being primarily derived from the action of the solids upon it, the chief spring or rise of motion is in the solids.[15] Physicians such as Fuller believed humours would 'partake' of the benefit gained from exercising the fibres, because ultimately, the fibres were in charge of circulating the fluids through the body.[16] As a result, medical literature, including the popular guides to health, placed great emphasis on the maintenance of the elasticity and strength of the fibres of the body in their advice on how to live long and healthily.

While many seventeenth-century texts suggested that exercise was beneficial, but optional, eighteenth-century physicians argued that exercise was *vital* to health. Exercise had long been considered to be one way in which the *excreta* could be increased to balance with the *ingesta*, but certainly not the only way. An increased excretion could be had by the use of various medicines, such as diuretics, and the heating aspect derived from exercise could also be had by other methods. Quincy suggested that inducing a fever might beget the same benefit, for example.[17] Additionally, simply adjusting one's diet to decrease the *ingesta* served the same purpose of preventing plethora. The idea that a moderate diet could be sufficient to ensure a good balance between *ingesta* and *excreta*, and therefore, health can still be found in many seventeenth-century texts, such as in the accounts of famous dieters Luigi Cornaro (1485–1566) and Leonardus

[14] Bernard Lynch, *A guide to health through the various stages of life* (London: Printed for the author, 1744), 273, Eighteenth Century Collections Online.

[15] Ibid., 272–274.

[16] Fuller, *Medicina Gymnastica*, preface.

[17] Quincy, *Medicina Statica* (version 1737), 54–55.

Lessius (1554–1623).[18] In his account of his lifestyle, Cornaro emphasised a moderate diet over all other means of obtaining and maintaining health. Lessius argued along the same lines as Cornaro, stating that 'Sobriety' in diet '[prevents] crudity of humours [and] consumes the superfluous humours, and that much more safely than bodily exercise doth'.[19] Since diet could prevent the very issues that exercise would cure, Lessius argued that for those who 'ordinarily to such as lead temperate and sober lives … there is no need of long walks, or long continued exercises, whereby much time is wasted and lost'.[20] Cornaro and Lessius did not deny that there were certain health benefits to exercise but seem to have considered it unnecessary unless you were making mistakes in your diet.

These arguments that it was diet above all when it came to health were occasionally still echoed in some eighteenth-century discussions of health. Essayist Joseph Addison (1672–1719), for example, stated in *The Spectator* in 1711: 'If exercise throws off all superfluities, temperance prevents them; if exercise clears the vessels, temperance neither satiates or overstrains them'.[21] However, the idea that a moderate diet could bypass the necessity of exercise seems to have all but disappeared from eighteenth-century health guides. In 1744, Bernard Lynch assured his readers that 'there is an *absolute necessity* for labour and exercise, to preserve the body any time in a due state to maintain health and prolong life', and both he and other physicians such as Cheyne warned with great urgency that exercise under no circumstance should be neglected, for 'without due exercise, the juices will thicken, the joints will stiffen, the nerves will relax, and chronical distempers and crazy old age must ensue'.[22] Unlike Cornaro and Lessius,

[18] These accounts were translated into English and published together as *The Temperate Man, or the Right Way of Preserving Life and Health … In Three Treatises, the first written by the learned Leonardus Lessius, The second by Lodowick Cornaro, a Noble Gentleman of Venice, the Third by a famous Italian.* (London, printed by J.R. for John Starkey, 1678); and *Hygiasticon; Or, A Treatise of the Means of Health and Long Life, Written originally in Latin by Leonardus Lessius, Now rendered into English by T. Smith, Whereunto is annexed, Cornaro's Treatise of the Benefits of a Sober Life* (London: Printed for the Author and sold by Charles Hitch, 1742).

[19] Lessius, in *The Temperate Man, or the Right Way of Preserving Life and Health*, 68.

[20] Ibid., 70.

[21] Joseph Addison, 'No. 195'', *The Spectator*, 3, 195 (October 13, 1711), 98–99. https://archive.org/details/in.ernet.dli.2015.502846/page/n107/mode/2up.

[22] Lynch, *A guide to health through the various stages of life*, 286, emphasis my own; George Cheyne, *An Essay of health and long life* (London: Printed for George Strahan, 1724), 90–91, Eighteenth Century Collections Online.

Fuller believed that the health benefits brought by exercise were unique to exercise alone, and so exercise could not be substituted with other non-naturals or internal medicines. This was because exercise, unlike the other non-naturals, directly affected the solids of the body by giving a greater tension to the fibres, thus increasing their elasticity. Some seventeenth-century texts had commented on the positive effects of exercise on the solids, but it was not until physicians had shifted their focus from the health of the fluids to the health of the solids that this health benefit became as relevant. Writing around the turn of the century, Fuller lamented that people had so far not placed much value on exercise and the health benefits it brought.[23] This shift affected advice on all the non-naturals but seems to have had especially significant effects on exercise advice, something that was also noted by Hisao Ishizuka, Edward Kennaway, and Rina Knoeff.[24]

This effect of augmenting the strength of the fibres could not be had as effectively from any other non-natural or medicine, although there were some medicines circulating the market that claimed to strengthen the solids. These alternative methods were by and large discredited by physicians in favour of exercise, however. 'How ridiculous is it', Fuller posed, 'for a person to expect to be perfectly restor'd to a firm habit of Body by internal Alternatives, and Methods ... which ... can never restore the Tone of the solids'.[25] Instead of relying on medicines to restore the solids, 'the Solids must be made to strengthen themselves, and recover their Spring by frequent endeavours', such as through exercise, Fuller concluded.[26] Cheyne also believed exercise was the only non-natural which could keep the solids of the body in a healthy state. 'Nay, the joint power of warm air and light food cannot supply the place of exercise in keeping the joints pliant and moveable, and preserving them from growing rusty and stiff', he argued.[27] 'Tho food may, in a great measure, prevent the thickening of fluids', Cheyne explained, 'yet it cannot do it sufficiently without exercise', as diet did not affect the solids like exercise did: 'for that purpose exercise is

[23] Fuller, *Medicina Gymnastica*, preface.

[24] Hisao Ishizuka, 'Fibre body': the concept of fibre in eighteenth-century medicine, c.1700–40", *Medical history*, 56,4 (2012), 575. doi:10.1017/mdh.2012.74; Kennaway and Knoeff, '"For it is the debilitating fibres that exercise restores", 108–111.

[25] Fuller, *Medicina Gymnastica*, p. 35.

[26] Ibid., 36.

[27] Cheyne, *An Essay of health and long life*, 91.

absolutely necessary'.[28] While exercise, as one of the six non-naturals, had always been considered important to health, the onset of iatromechanism in the early eighteenth century lifted its status considerably so that it was now considered an absolute necessity to the health of the body.

Not only did exercise advice enjoy greater attention than before, but the advice itself also changed in several ways. According to mechanical physicians, all the individual fibres of the entire body had to be exercised, preferably simultaneously. Physicians therefore advised exercises which affected all parts of the body equally.[29] Furthermore, in order to affect the minute fibres, the exercises themselves had to be minute. Physicians recommended exercises which sent 'vibrations' through the body small enough to reach even the most minuscule fibres. To truly understand how different this is from modern ideas of what constitutes exercise, one only has to consider Cheyne's argument that vomiting could be considered a very effective form of exercise, for not only did vomits drive out noxious contents, but they also squeezed and compressed, motions which, 'by setting the whole System, and the most remote Fibre, in violent Action and Play', could supply the place of exercise and were therefore 'the most Universal Remedy'.[30] By Cheyne's own admission, these motions were quite 'violent' and therefore not suited to everybody, particularly not pregnant women as it could cause miscarriage. A similarly 'un-modern' but certainly more pleasant exercise recommendation came from Bernard Lynch in 1744, who offered that music, 'tho 'strickly speaking, may not be deem'd exercise', nevertheless affected the fibres of the body through minute vibrations or 'thrillings' in the air, and was therefore not only good exercise for the mind but likewise for the body.[31] These suggestions were rather exceptional, as both Cheyne and Lynch admitted, but they do nicely illustrate how physicians imagined exercise affected the body and how this differs from our own understandings of exercise and health.

Usually, health guides named horseback riding and walking as the most healthful of all exercises, on the premise that these exercises applied a 'universal shaking' to all the fibres of the body. 'The best exercise', Francis de Valagin (1724–1804) told his readers,

[28] Ibid.

[29] Tennis or football were recommended to those who specifically suffered from weak arms of legs. Lynch, *A guide to health through the various stages of life*, 287–288.

[30] Cheyne, *An essay on regimen*, v-vi.

[31] Lynch, *A guide to health through the various stages of life*, 284.

> is that which gives an equal motion to all the parts … for it puts all the muscles into action, it swells them, and thus accelerates the return of venal blood towards the heart; which stimulates it into a swifter contraction; whence the circulation of all the humours becomes quicker, and from the increased motion of the solids and the fluids, the digestion and assimilation of our food into healthy humours is greatly promoted.[32]

Conversely, exercises which only affected certain body parts were deemed less effective and were therefore not advised as much as riding, unless the specific circumstances of a patient called for it. For those people who, for health reasons or reasons of convenience, could not ride a horse, physicians recommended exercises which emulated the vibratory sensations of horseback riding. A coach or chariot was best for the additional benefit of being exposed to fresh air, but the indoor 'chamber-horse' could suffice to at least bring the benefit of the vibratory motion, if not the clean air. Cheyne recommended the chamber-horse to his exercise-shy friend, Samuel Richardson, on account of Richardson's 'flabby state of the solids'.[33] The chamber-horse was simply a wooden chair with a leather seat containing springs which allowed the user to bounce up and down, emulating the trotting of a horse. It is important to mention here that generally chamber-horses were recommended only to those who were unable to ride due to some illness or impairment—generally, it was believed important that exercise was undertaken in the fresh air. Physicians were, however, generally happy for people to exercise in whatever way they pleased, and other historians have already demonstrated that a wide range of sports and exercises were practised in the early modern period.[34] Nevertheless, health guides of this period made it clear that the most healthful exercise was that which targeted the fibres of the body universally, resulting in

[32] Francis De Valagin, *A treatise on diet, or, the management of human life; by physicians called the six non-naturals* (London: Printed for the author by J. and W. Oliver, 1768), 215, Eighteenth Century Collections Online.

[33] Cheyne to Richardson, May 13th, 1739, *The Letters of Doctor George Cheyne to Samuel Richardson*, 52.

[34] Alessandro Arcangeli, "Exercise for Women", in *Sports and Physical Exercise in Early Modern Culture*, edited by Rebekka Von Mallinckrodt and Angela Schattner (Abingdon: Routledge, 2016), 147–164; Dennis Brailsforth, "Sporting Days in Eighteenth Century England," *Journal of Sport History*, 9, 3 (Winter, 1982). 41–54; Mike Huggins, "Early Modern Sport," in *The Oxford Handbook of Sports History*, eds. Robert Edelman and Wayne Wilson (Oxford: Oxford University Press 2017), 113–128.

exercise recommendations that by our modern standard would perhaps seem rather low impact or indeed no exercise at all.

The idea that exercise was absolutely vital to health did not wane even as iatromechanism lost ground to competing schools of thought in the later eighteenth century. Over the course of the eighteenth century, authors of health guides disseminated warnings about the detrimental consequences of insufficient physical exercise with increasing frequency and greater urgency. According to Kennaway and Knoeff, physicians in the latter half of the eighteenth century continued to emphasise the vital role of the fibres on the functioning of the body.[35] In fact, Kennaway and Knoeff argue that the theory of the fibrous body caused late eighteenth-century physicians to recommend 'vigorous' exercises which focussed on 'strengthening' these fibres and were more vigorous in nature than exercises had been before the second half of the eighteenth century.[36] However, my own study of eighteenth-century exercise advice shows that many late eighteenth-century physicians continued to recommend 'moderate' exercises as opposed to 'vigorous' exercises, as we shall see in Chap. 5.

Although many warnings were motivated by iatromechanical concerns with the fibres, concerns with exercise levels continued to find expression in health literature long after mechanical theories lost ground to other schools of thought. Shifts in medical theory were not the only cause for the eighteenth-century focus on exercise as being vital to health maintenance. From the second half of the eighteenth century, physicians in health literature expressed considerable concerns with changing lifestyles as a consequence of urbanisation. It is from 1750 onwards especially that the term 'sedentarism' starts to appear in health guides. What exactly constituted a sedentary lifestyle is not often quantified, although Buchan gives an indication of where the line might have been drawn: 'a man shall not be hurt by sitting four or five hours a day, but if he is obliged to sit ten or twelve, he will soon contract disease'.[37] As we will discuss in the next section, changes in the nature of employment, leisure activities, and even travelling habits caused physicians to continue to emphasise the importance of exercising the fibres of the body.

[35] Kennaway and Knoeff, "For it is the debilitating fibres that exercise restores", 111–113.

[36] Ibid.

[37] William Buchan, *Domestic medicine; or, a treatise on the prevention and cure of dieases by regimen and simple medicines.* 2nd edition (London: Printed for W. Strahan, 1772), 52, Eighteenth Century Collections Online.

Exercising a Nation

Concerns about exercise and people's exercise needs continued to find expression throughout health advice in the second half of the eighteenth century, though there was a noticeable difference between exercise advice written in the early and mid-eighteenth century and that written by the close of the century. Where earlier warnings about the lack of exercise concern themselves with the exercising habits and state of health of individuals, later texts focus more on the exercise levels of the urban population as a whole and on the detrimental effects that a lack of exercise might have on the nation, rather than just the individual. This trend fits in with similar, though less pronounced developments in dietetic advice, where late eighteenth- and nineteenth-century texts increasingly explored the dietary needs of the health of the population as a whole and not just the needs of the individual.[38] A particularly striking difference between early and late eighteenth-century discussions about exercise is physicians' use of terms related to exercise. Early physicians often spoke of 'idleness' when referring to a person's lack of exercise, while late eighteenth-century physicians used the term 'sedentarism'. While the term 'sedentarism' does appear in some early eighteenth-century texts, it nevertheless appears much more frequently in late eighteenth-century discussions. Similarly, the term 'idleness' does not disappear altogether from late eighteenth-century texts but does seem less in use by this time. Though both indicate insufficient exercise, these terms are sufficiently different to warrant some unpacking. 'Idleness' was traditionally described as an individual trait or habit. When authors of medical texts discussed idleness, they entreated that the idle individual change his or her habits. They did not usually suggest that idleness was an external issue. Sedentarism, however, was often ascribed to societal issues, such as occupations or living circumstances, rather than individual shortcomings.

[38] As discussed in the previous chapter, especially pp. 78–79. For further discussion, see Maria Pia Donato, "The afterlife of the non-naturals in early eighteenth-century Hippocratism: from the healthy individual to the healthy population," in *Conserving Health in Early Modern Europe: Bodies and environments in Italy and England,* edited by Sandra Cavallo and Tessa Storey (Manchester: Manchester University Press 2017); Rebecca Earle, "The Political Economy of Nutrition in the Eighteenth century," *Past and Present,* 242, 1 (Feb. 2019), 83, and Emma Spary, "Food," in *A Cultural History of Medicine: In the Enlightenment* ed. Lisa Wynne Smith (London: Bloomsbury 2021), 29–50.

Late eighteenth-century health guides discussed not only the personal consequences of insufficient exercise but also how sedentarism might affect society more widely. It was suggested, for example, that sedentary men were less able to perform well in their employments. A sedentary life made men 'weak, sickly, puny, and effeminate' and, consequently, 'useless to society', one physician argued.[39] For women, a lack of exercise was thought to lead to detrimental effects on their reproductive abilities. Scottish physician William Buchan (1729–1805) warned that 'want of exercise not only relaxes the body, but induces a plethora, or too great a fulness of the vessels, which are the two principal causes of abortion', and physicians therefore recommended that pregnant women should generally continue their usual exercises.[40] This idea was not new to the eighteenth century, as English midwife Jane Sharp (1641–1671), argued that women living in cities who therefore had a less active lifestyle were less likely to conceive than country women, for whom physical work was part of their lifestyle. She recommended:

> Moderate exercise, for idleness is a great enemy to conception; and that may be reason that so many City Dames have few children & if they have any, they are commonly sickly and short lived; it is not so with Country women who are always working; they usually have many children; and they are lusty and strong, for moderate labour raises natural heat, revives the spirits, helps digestion, opens the pores, and wastes excrements, comforts all the parts, and strengthened the senses and spirits, helps nature in all her faculties, and that is the way to have … many children.[41]

Even when conception was successful, parents' sedentary lifestyles could still negatively affect their offspring, physicians argued. They believed that constitutional strengths and weaknesses could be passed on from parents to children. 'A delicate female, brought up within doors, an utter stranger to exercise and open air, who lives on tea and other slops, may bring a child into this world, but it will hardly be fit to live', Buchan

[39] *The universal family physician, and surgeon. Containing a familiar and accurate description of the symptoms of every disorder incident to Mankind; together with their gradual Progress, and method of cure* (London, 1798), p. 112; De Valagin, *A treatise on diet*, 216.

[40] Buchan, *Domestic Medicine*, 336.

[41] Jane Sharp, *The midwives book, or, The whole art of midwifery discovered.* 2nd edition (London: Printed for Simon Miller, 1671), 178, Early English Books Online.

argued.[42] He continued: 'When indolence and sedentary employments prevent two-thirds of mankind from either taking sufficient exercise themselves or giving it to their children, what do we expect but diseases and deformity among their offspring?'[43]

In their discussions, physicians seemed primarily to blame sedentary occupations for the insufficient levels of exercise so prominent especially among the urban population. The concern that being seated all day as part of one's occupation could be detrimental to one's health was not new to the late eighteenth century. Already at the end of the seventeenth century, Italian physician Bernardino Ramazzini wrote *De Morbis Artificum Diatriba*, which was first published in Latin in 1700, and then translated into English in 1705 under the title *A Treatise on the Diseases of Tradesmen*.[44] This work included a chapter concerning the 'diseases of such tradesmen as sit much', drawing attention to sections of society which were hitherto not mentioned in any discussions on exercise or idleness. Physicians of the seventeenth and early eighteenth centuries were well aware that some occupations involved what they considered to be an unhealthy amount of sitting. Usually these were understood to be students or 'men of letters', which constituted a much narrower section of society than the groups Ramazzini warned to be in danger of the consequences of sitting too much. But while Ramazzini's treatise may have contributed to later discussions about sedentarism, his initial warnings about sedentary occupations did not carry the same level of urgency as later warnings did.

William Buchan was one of the first in Britain to extensively discuss the dangers of sedentary occupations, dedicating a chapter to 'the laborious, the sedentary, and the studious' in his *Domestic Medicine*, the first edition of which was published in 1769.[45] Buchan had claimed that '[t]hose that could be reckoned sedentary' now comprehended 'the far greater part of the species' and that '[a]lmost the whole female world, and in manufacturing countries, the major part of the males, may be reckoned sedentary'.[46] Bryan Cornwell went further still and warned that 'the inhabitants of this country seem to be in some danger of losing the use of their limbs', which

[42] Buchan, *Domestic Medicine*, 8.

[43] Ibid., 24.

[44] Bernardino Ramazzini, *A treatise of the diseases of tradesmen, shewing the various influence of particular trades upon the state of health* (London: Printed for Andrew Bell et al., 1705), Eighteenth Century Collections Online.

[45] Buchan, *Domestic Medicine*.

[46] Ibid., 51.

illustrates well how widespread and severe medical writers of this period believed sedentarism to be.[47] J.D. Adair included in his definition of the term 'sedentary' not only studious men but also 'inactive women of every station, shop-keepers, clerks, and men who work in a sitting posture'.[48] In the late eighteenth century and well into the nineteenth century, sedentary employments and their health detriments were addressed in medical treatises by Bryan Cornwell (1784), A.F.M. Willich (1800), Joshua Webster (1804, p. 157), Joseph Taylor (1818), T.J. Graham (1827), and Christopher Hufeland (1829), and came up in multiple letters received and written by William Cullen throughout the 1770s and 1780s.[49] By the early nineteenth century, T.J. Graham explained that he had allotted a larger space to exercise in his treatise than to any other subject, 'on account of its general neglect'. He lamented the 'debauchery of inaction' that had spread itself 'extensively throughout this nation' and produced an 'alarming increase' of chronic diseases.[50]

Edinburgh-based physician William Cullen (1710–1790) received many letters from other doctors worried about their sedentary patients. Robert Hunter's physician, for example, had said that Robert's corpulency had come about as a result of his sedentary employment, where he before had been 'healthy all his life long'.[51] Mr. Murray's physician also believed Murray's constitution to have suffered as a result of 'a sedentary life & much application to business'.[52] One of Cullen's own patients, who had been plagued by weak and obstructed bowels, was also told by Cullen that

[47] Bryan Cornwell, *The domestic physician; or, guardian of health* (London: J. Murray, J. Bew, L. Davis, 1784), 69, Eighteenth Century Collections Online.

[48] J.M. Adair, *An Essay on Diet and Regimen as indispensable to the recovery and enjoyment of firm health.* 2nd Edition. (London, 1804, second edition), xii

[49] Cornwell, *The Domestic Physician*; Anthony Willich, *Lectures on diet and regimen: being a systematic inquiry into the most rational means of preserving health and prolonging life* (London: Printed for T.N. Longman and O. Rees, 1799), Eighteenth Century Collections Online; Joshua Webster, *Practical observations on the preservation of health and instructions for the preventions and cure of human maladies* (London: J. Slee, 1804), https://wellcomecollection.org/works/rsktaxjp; Joseph Taylor, *Annals of health and long life: with observations on regimen and diet, so necessary to the preservation of life* (London: Printed for Effingham Wilson, 1818), https://wellcomecollection.org/works/vexhyptv.

[50] Graham, *Sure Methods of Improving Health and Prolonging Life*, vii-viii.

[51] Unknown to William Cullen, Reply headed 'Case of Robert Hunter', 18 July 1770 (Doc. ID:6247), "The Consultation letters of William Cullen", Royal College of Physicians at Edinburg. accessible online via *The Cullen Project*, <http://www.cullenproject.ac.uk>.

[52] William Cullen to Unknown, 'Reply for Mr Peter Murray', 14 May 1770 (Doc. ID:89).

this was a consequence of his 'sedentary life with intense application to business'.[53] If a sedentary profession was the problem or part of the problem, the patient in question would be advised to 'lay aside all sedentary business & to get some health & strength' and take some exercise.[54] Patients themselves also seemed aware that their sedentary work could pose a threat to their health. Cullen's notes included a copy of a letter drawn up by a patient, named Mr. Rudd. Mr. Rudd described himself as having lived a 'sedentary and studious life' and admitted he had never taken sufficient exercise.[55] To write this, Mr. Rudd must have believed a lack of exercise to have been a factor in his ailments, and other patients too felt the need to include their exercise levels or lack thereof when writing to their physicians. A surveyor, named John Buchanan, wrote: 'Am now aged about 35, have a wife & family my occupation is sometimes sedentary, at other times I am pretty much abroad and get moderate exercise'.[56] One Mr. J.C., a student, had a range of complaints that he himself ascribed to 'intense thinking on Metaphysical & Mathematical subjects & the very sedentary life such employment requir'd'.[57] Letters such as these demonstrate the extent to which both physicians and patients believed sedentarism to play a part in their ill health.

Unfortunately, Cullen and his correspondents rarely mention what sort of business the patient in question is involved in, but when they do, these are activities relating overwhelmingly to those from the upper echelons of society. This probably has more to do with Cullen's eminence as a physician and consequently high fees rather than with a lack of concern from those of less elite backgrounds, but this is difficult to establish with any certainty. There is therefore little evidence to ascertain whether and how tailors, shoemakers, and shopkeepers balanced their sedentary work with any physical exercise. Buchan, however, does shed some light on these lower social ranks, writing that 'sedentary artificers are generally allowed too little time for exercise', suggesting limitations in leisure time could be as much of a constraint in getting sufficient exercise as lack of funds. Aside

[53] William Cullen to W.F., 'Reply for Mr W. F.', August 1764 (Doc. ID:13).

[54] Ebenezer Jacob to William Cullen 'Letter from Dr Ebenezer Jacob', 1 May 1779 (Doc. ID:1664); William Cullen to Mrs Pratt, 'Reply' 22 May 1764 (Doc. ID:762).

[55] Mr Rudd to William Cullen, 'Copy of a letter by Mr Rudd', 1 December 1783 (Doc. ID:2377).

[56] Unknown to William Cullen, 'Letter from John Buchanan', 17 February 1779 (Doc. ID:1629).

[57] Unknown to William Cullen, 'Case of J.C.', 16 November 1778 (Doc. ID: 1587).

from this constraint, however, Buchan was also under the impression that sedentary artificers did not care enough to use what little time they had on exercising their bodies. '[Y]et short as it is', Buchan continued, 'they seldom employ it properly … a journeyman tailor, instead of walking abroad for exercise and fresh air, at his hours of leisure, chooses often to spend them in a public house, or in playing at some sedentary game'.[58] While Buchan's choice of words seems to suggest that the tailor lacked the motivation, rather than the knowledge to make a healthier choice, it is important to keep in mind that such 'observations' regarding the labouring classes may have a strong basis in prejudice rather than reality. Most physicians, after all, came from the middle and upper classes and seemed of the opinion that the 'rougher' sorts did not have the sensibility to look after their own health.[59]

It appears that late eighteenth-century physicians were arguing that widespread sedentarism was a contemporary and a societal issue. This shines through especially in this excerpt from Graham's *Sure Methods of Improving Health*:

> A strict and spare diet is at this time … necessary; but this is an unnatural state, imposed on man in the present day, because his physical energies are greatly impaired by sedentary and artificial habits, and his stomach, consequently, is not able to bear food in the quantity and quality that our forefathers fed upon … It is the "debauchery of inaction" and not of repletion, which has spread itself so extensively throughout this nation, and produces so alarming an increase of chronic diseases; and it is only by a return to active bodily exertion that we can hope to rise above this unnatural condition.[60]

Graham's assertion that this unnatural state of inaction is *imposed on man* rather than a failure of man is especially indicative of the difference between early and late eighteenth-century discussions about 'idleness' and 'sedentarism'. While the former term was used to reprimand individual readers for their poor habits and willpower, the latter term was employed to call readers' attention to developments which were happening around them, rather than within them.

[58] Buchan, *Domestic Medicine*, 60.

[59] Michael Stolberg, "A Woman Down to Her Bones: The Anatomy of Sexual Difference in the Sixteenth and Seventeenth Centuries", *Isis*, 94, 2 (June 2003), 297.

[60] Graham, *Sure Methods of Improving Health*, Vii-Viii.

Sedentarism was therefore, by the late eighteenth century, no longer considered to be an issue limited to the wealthy elite. Physicians were explicit in their concerns that it had become widespread to the extent that it posed a significant threat not just to the sedentary individual but indeed to the nation as a whole. 'In my opinion', wrote Graham in 1827, 'the chief cause of indigestion, bilious and liver complaints, is a sedentary mode of living, which is now carried to a great extent by all classes of persons in Great Britain, who rise above the lower orders'.[61] As shown above, Buchan believed the same, stating that a sedentary lifestyle was now common to 'the greater part of the species', and especially so among women, and people working in the manufacturing industry.[62]

Sedentarism in the City

While it is evident that many physicians argued sedentary occupations to be the primary cause behind what Adair termed the 'debauchery of Inaction', what is less clear is why physicians started to concern themselves with sedentary professions in the first place. There is a possibility that Buchan, Graham, and Adair witnessed a particularly rapid transformation of the nature of occupations, though it seems an unsatisfying explanation. Ramazzini called attention to the many different sedentary occupations and their detrimental health consequences a hundred or so years before Buchan, Graham, Adair, and many others did. In other words, sedentary professions were nothing new, and there has been no indication that the physical nature of work changed significantly in the late eighteenth century.

Nevertheless, physicians appeared to be under the impression that exercise levels had changed recently and significantly. Aside from the sedentary nature of many occupations, they also identified other concerning trends which contributed to this 'debauchery of inaction'. Most, if not all, of these concerns were explicitly linked to life in large towns and cities. Late eighteenth-century physician Robert Priestly considered the cases of bilious diseases that he witnessed to be the consequence of urbanisation:

> The man destined to pass his days in the heart of large towns or cities, and employed in affairs of importance, cannot move about so much as he, who

[61] Graham, *Sure Methods of Improving Health*, 181.
[62] Buchan, *Domestic Medicine*, 51.

> resides in the country. He frequently finds, that his sedentary manner of living is detrimental to him … he is more to be pitied than blamed. [Many] … are arrived to such a degree of indolence, that they not only renounce [exercise] but fear it; and that means, they deprive themselves of the most agreeable and sure way of diminishing the danger of many ills which their conduct produces … and all persons who have time to employ themselves as they please, should lay it down as a rule, to exercise themselves every day, at least two hours, in walking or riding.[63]

Buchan argued that the diseases which were common among people living sedentary lifestyles (such as scopula, consumption, hysterics, and nervous diseases) were 'little known among people of active employments without doors'.[64] Buchan continued: 'The rickets … never appeared in Britain till manufacturers began to flourish … and people … left the country to follow sedentary employments in great towns'.[65] De Valagin similarly expressed his conviction that urbanisation and industrialisation were to blame for changes in lifestyles and the consequent health issues. Moving from the countryside to the city involved a shift away from the physical activities which were inherent to rural life. He argued:

> The nearer we draw towards the Metropolis, London, the Center of Arts and Sciences, the Center of Trades and Commerce, the Center of Plenty and Riches, and we may say also, the Center of Luxury and Voluptuousness, the more we find this Subversion of the Manners of our Ancestors to take place. In former days, out Farmers in the Country taught their Sons to follow the Plough, and the Daughters to Spin, to milk the Cows, and to make Butter and Cheese, all healthful Exercises.[66]

This is not to say that sedentary lifestyles were never experienced by people living in the countryside, but it is clear that physicians were convinced that sedentarism was an issue especially in urban environments.

Concerns were not limited to men of sedentary professions but applied to all who spent such an amount of time in the day being seated. This

[63] Robert Priestley, *A few interesting remarks on bilious disorders, particularly the jaundice, and yellow fever, &c. Highly important to persons of fashion, the sedentary, and those who inhabit or frequent tropical climates; recommending the anti-bilious powder, a certain and experienced remedy* (London: Printed by Thomas Gill, 1798), 64.

[64] Buchan, *Domestic Medicine*, 60

[65] Ibid., 44.

[66] De Valagin, *A treatise on diet*, 8–9.

included women whom physicians believed to be spending much of their day sitting down—only 'a very small proportion of females lead a very active and laborious life', claimed Adair.[67] Part of this was by design. 'Women in all civilized nations, have the management of domestic affairs, and it is very proper they should, as nature has made them less fit for the more active and laborious employments ... [women] are fitter for every kind of business which does not require strength', Buchan argued, using medical evidence of the weaker female body to support the culturally determined division of labour.[68] Buchan was not the only nor the first to use this line of reasoning for arguing that women were predisposed to live a sedentary life (and suffer the consequences of such a lifestyle). Historians such as Thomas Laqueur have shown how medical depictions of women's bodies were often strongly informed by cultural ideas about women's role in society.[69] Michael Stolberg, for example, found that anatomical drawings of the early modern period depict women with flatter sitting bones which were to reflect their naturally being more suited to sedentary lifestyles.[70]

It is possible that women's roles being tied to the household may have restricted the amount of exercise they had on a day-to-day basis in comparison to men. Certain household tasks done by women may have been more sedentary in nature than the tasks performed by men. Famed Swiss physician Samuel-Auguste Tissot (1728–1797) expressed that 'women[s] ... light sedentary work affords them less exercise and motion than the more active occupations of men'.[71] But domesticity did not imply sedentarism per se. Whether women's work was light depended primarily on their socio-economic status. Running a household could also require performing various physically taxing tasks, ranging from feeding and tending to livestock, fetching water, and washing laundry, to name a few.[72]

[67] Adair, *Essay on diet and regimen*, 100–101.

[68] Buchan, *Domestic Medicine*, 83 and 330–331.

[69] Thomas Laqueur. *Making Sex_ Body and Gender from the Greeks to Freud*. London: Harvard University Press, 1992.

[70] Stolberg, "A woman down to her bones", 281.

[71] Samuel Auguste Tissot, *Advise to the people in general, with regard to their health ... Translated from the French edition of Dr. Tissot's Avis au people &c. Printed at Lyons, with all the notes in the former English editions, and a few additional ones, by J. Kirkpatrick* (Dublin: Printed J. Hoey Senior, 1765). 354, Wellcome Collection Online.

[72] Tim Reinke-Williams, *Women, Work and Sociability in Early Modern London*. London: Palgrave Macmillan, 2014. DOI: https://doi.org/10.1057/9781137372109, 47.

Women of the upper classes, however, especially those living in London, were often accused of neglecting their duties as housewives, delegating the more arduous tasks to servants, thus taking away hours of physical work, and indulging themselves in time spent in leisure.[73] It was this kind of lifestyle that had physicians concerned, as these women, by shirking such tasks, robbed themselves of much necessary physical exercise.[74]

It was not always the lack of housework that worried physicians: women's employments were also part of the problem. It was believed that especially city women were involved in work that required sitting for long periods of time. For example, Buchan ascribed Miliary fever said to be found in women more often than men, especially to 'the delicate and indolent, who, neglecting exercise, keep continually within doors'.[75] However, it was also 'extremely fatal to … women in manufacturing towns, who, in order to assist their husbands, sit close within doors for almost the whole of their time. But among women who are active and laborious, who live in the country, and take sufficient exercise without doors, this disease is very little known'.[76] While the running of the household was unquestionably a woman's domain, women in this period, especially those living in urban areas, were not confined to the household and many participated in their husband's business. Poorer women might support their husbands by retailing goods in the streets while those of the middling sorts might run stalls or shops.[77] Working in a shop or stall might require hours of sitting, as might producing goods for sale, as we can see from this excerpt concerning a wife who, in helping her husband, 'rose very early in the morning and sat until it was very late at night, she not having many times in a whole day's space together so much as rose from the stool she sat making periwigs'.[78] As Buchan's statement suggested, the problem of overly sedentary work was believed to almost exclusively apply to women living in towns or cities. Tissot similarly argued that 'country girls, who are generally more accustomed to such hardy work and exercise as county men, are less subject to … complaints than women who live in cities'.[79] As with housework, to which extent work was physically straining

[73] Ibid., 56–57.
[74] Tissot, *Advice to the People*, 354.
[75] Buchan, *Domestic Medicine*, 144.
[76] Ibid., 148–149.
[77] Reinke-Williams, *Women, Work and Sociability in Early Modern London*, 103–104.
[78] Ibid., 51.
[79] Tissot, Advice to the People, 256 and 356.

or sedentary depended much on socio-economic factors, especially on the nature of the business women were involved in.

In addition, physicians were concerned with the number of hours children spent being seated at their school desks. Exercise advice concerning children could be found in literature which focussed on child health and child rearing, the authors often stressing the importance of physical education. This was especially the case towards the end of the period. These texts generally promoted gymnastic sports for children and advised on how to combine long school days with physical education.[80] The authors of these texts were concerned about the sedentary nature of education, and the consequences that so much time spent seated had on the health of English children. '[T]he poor child is fixed to a seat seven or eight hours a-day, which time ought to be spent in exercise and diversions. Sitting so long cannot fail to produce the worst effects upon the body', Buchan argued.[81] 'Let a child or a youth daily exercise his vigour in the open air, by running, jumping &c; and I engage he will never fall into the detestable vice of onanism. It is peculiar to a sedentary education, in schools and other seminaries, where exercise is confined to half-hours', Hufeland warned.[82] De Valagin similarly expressed that 'now' education was 'very different, they are sent to Boarding Schools, and from their tenderest Years accustomed to the Sweets of luxurious Life'.[83]

Physicians started to provide advice to parents, and especially mothers, on how to ensure that their children had sufficient exercise at home. Buchan especially argued that the foundations of health were laid by mothers, who 'have it very much in their power to make men healthy, or valetudinary' (a term often used to describe people who were generally sickly or weak of body).[84] Sedentarism was believed to be more prevalent among girls than boys. T.J. Graham noted that especially elite girls were not provided with the opportunity to exercise and perhaps even prevented from doing so. Graham cited surgeon Sir Ashley Cooper, who noted that boys were less likely to suffer from scrofula than girls because the former

[80] Christopher Hufeland, *The Art of Prolonging Human Life* (London: Published by Simkin and Marshall, Stationer's Court, 1829), 230–233.

[81] Buchan, *Domestic Medicine*, 45–46.

[82] Hufeland, *The Art of Prolonging Human Life*, 233.

[83] De Valagin, *A treatise on diet*, 9.

[84] Buchan, *Domestic Medicine*, 1–5.

group took exercise, which girls were 'not allowed'.[85] 'Would mothers, instead of having their daughters instructed in many trifling accomplishments, employ them in plain work and house-wifery, and allow them sufficient exercise in the open air, they would both make them more healthy mothers, and more useful members of society', Buchan argued.[86] The poor were not exempt from Buchan's criticism. 'Mothers of the poorer sort think they are great gainers by making their children lie or sit while they themselves work', he perceived, but 'by neglecting to give their children exercise, they are obliged to keep them a long time before they can do any thing for themselves, and to spend more on medicine than would have paid for proper care'.[87] Unlike the wealthier mothers, however, the poorer mothers were understood to have less power to control the exercise levels of their children: 'poverty often obliges them to neglect their offspring in order to procure the necessities of life. When this is the case … it becomes the duty of the public to assist them'.[88] Although childhood sedentarism was found to occur in the countryside as well, it was nevertheless usually attributed to urban households especially.

The posture of the sedentary body was considered almost as problematic as the lack of motion. It appears that the reasoning behind concerns with poor posture was quite similar to concerns with lack of movement, however. Both threatened to obstruct the circulation of the bodily fluids, physicians explained. Francis Spilsbury argued that '[A] bending position obstructs all the vital motions, and … bring[s] on an universal relaxation of the solids'.[89] The result of this, according to Spilsbury, was a stagnation of the blood and obstruction of the perspiration, 'from whence proceed the scab, ulcerous sores, foul blotches, and other cutaneous diseases, so common among sedentary artificers'.[90] 'It is impossible that the man who sits from morning till night should either digest his food, or have any of the secretions in due quantity', Spilsbury concluded.[91]

[85] Graham, citing sir Astley cooper, *Sure Methods of Improving Health and Prolonging Life*, 183.

[86] Buchan, *Domestic Medicine*, 47.

[87] Ibid., 45.

[88] Ibid.

[89] Francis Spilsbury, *Free Observations on the Scurvy, Gout, Diet, and Remedy* (London: W. Gillman, 1788), 93, Eighteenth Century Collections Online.

[90] Ibid.

[91] Ibid., 95.

Graham had seen cases of curvature of the spine and growing out of the shoulder 'very much increased of late years'. Rarely seen among boys or young men but frequently found 'among girls and young women, especially in the more respectable classes of society ... debility is their chief cause, and exercise their only remedy'.[92] It was so common, he warned, that it ought to be 'a source of alarm to almost every respectable family', adding that such deformities were seldom seen amongst the poor and were almost entirely confined to girls.[93] Explaining why spinal deformation was becoming so common in young girls, Graham stated that the answer lay predominantly in the 'defects of the modern system of physical education'.[94] Graham warned that girls should avoid sedentary employment and that they should not be kept in school for more than 'a very moderate period' and that teaching sessions should never be extended beyond two hours at a time.[95] These periods of learning were to be broken up by time allocated for outdoor exercise to strengthen the muscles, which, consequently, would be better equipped to support the spine.[96] 'Parents should know and consider', Graham argued, 'that under a deficiency of exercise, the bones, especially of young children, invariably become soft and yielding, and that in endeavouring to impart solidity and strength to them, nothing can supply the place of exercise in the open air'.[97] The hours spent being seated, like the hours spent in rest, were only considered to be a problem to health when done in a manner which was not then balanced with time spent in exercise. Though sedentary occupations and education were not new to this period, it appears that physicians of the late eighteenth century were particularly of the opinion that that balance had been lost.

Sedentary work and school days were all the more problematic as these seemed to have come in tandem with a decline in physical activity outside of these hours. Physicians looked on in anguish at what they believed to be the increasing number of people who chose to travel by coach or chaise rather than by horseback or on foot. Originally, women predominated among passengers, as the use of coaches was looked down upon as

[92] Graham, *Sure and certain methods*, 192.
[93] Ibid.
[94] Ibid., 192–194.
[95] Ibid., 193.
[96] Ibid., 194.
[97] Ibid., 195.

unmanly.[98] Over the course of the seventeenth and eighteenth centuries, however, this was starting to change. Peter Edwards found that in the period 1650–1750 estate and household records are full of references to coaches, showing people were less opposed to these methods of travel than they had been before.[99] Physicians also remarked that coaches were now everywhere: 'the present fashion has made it quite ungenteel to employ our limbs', observed de Valagin, and Cornwell, as noted before, was convinced that the problem was so extensive that the inhabitants of Britain were 'in some danger of losing the use of their limbs'.[100] How different this was from the past: 'The old English certainly were intemperate livers, but in those days, if a man or woman was obliged to go a little way, it was on foot; if to a greater distance, it was on horseback … there was abundant exercise taken in the open air', Graham lamented.[101] This trend was, of course, restricted to those who could afford a coach, but physicians certainly seemed under the impression that growing numbers of people were in a position to do so.

That the city was the putative centre of this 'debauchery of inaction' was not always blamed on its inhabitants. It was generally understood that living in the city made exercise impractical and sometimes near impossible. Joseph Addison, for example, mentioned how he struggled to exercise in town and his solution to this problem:

> There is no kind of exercise which I would so recommend to my readers of both sexes as this of riding, … For my own part, when I am in town, for want of these opportunities, I exercise myself an hour every morning upon a dumb bell that is placed in the corner of my room, and it pleases me the more because it does every thing I require of it in the most profound silence. My landlady and her daughter are so well acquainted with my hours of exercise, that they never come into my room to disturb me while I am ringing.[102]

[98] Peter Edwards, *Horse and Man in Early Modern England* (London: Bloomsbury, 2007), 215–216. http://dx.doi.org.uoelibrary.idm.oclc.org/10.5040/9781474210072.ch-005.

[99] Edwards, *Horse and Man in Early Modern England*, 215–216.

[100] De Valagin, *A treatise on diet*, 14; Bryan Cornwell, *The domestic physician: or, guardian of health* (London: J. Murray, J. Bew, L. Davis, 1784), 69. Eighteenth Century Collections Online.

[101] Graham, *Sure and certain methods*, 177–181.

[102] Joseph Addison, "No.115", *The Spectator*, eds. Joseph Addison and Richard Steele, 1, 115 (July 12, 1711), 120.https://archive.org/details/in.ernet.dli.2015.39152/page/n437/mode/2up?q=dumb+bell.

Addison did not say why he lacked the opportunity to ride in London, but it may have been a matter of simply lacking the space to do so. London and its suburbs contained no fewer than six thousand coaches, Edwards found, clogging up the streets and bringing traffic to standstills.[103] The congested city streets may have made riding or even walking for exercise difficult while the polluted air would have made it decidedly less healthy than exercise taken in the countryside. Addison was able to afford dumbbells and had the space to use them, but those in less privileged positions might not have had the opportunities to exercise in the city.

These issues of space and costs may have led to large sections of the population in towns and cities being deprived of sufficient bodily exercise. One excerpt from De Valagin seems to suggest that exercise was increasingly difficult to be had in London in the late eighteenth century following the loss of one particularly popular place to exercise. De Valagin explained:

> It is the Usefulness of the Diversity of Exercises, that has always induced large and well regulated Cities to allot particular Play-Grounds, or Places for public Exercise, where Labourers and People who work at particular Trades, might at certain Hours of Recreation amuse themselves ... in order to prevent those Diseases which might arise from the particular Posture required in their Business. ... The Necessity of such a Place as Moorfields, in so populous a City as this, is therefore very obvious; and those who advised the Building of Streets upon that spot, have perhaps not maturely considered the Advantages which accrue to the Health of the immense Number of People who resort there every day.[104]

Limited time and space or not, physicians believed that individuals had to take responsibility for how they spent what little leisure time they had. '[A]ctive diversions have lost ground of late, and those of a sedentary kind seem to prevail', Buchan observed and argued that 'every thing that induces people to sit still ought to be avoided'.[105] Buchan praised a gardening scheme in the town of Sheffield, where 'there is hardly a journeyman cutler who does not possess a piece of ground, which he cultivates as a garden. This practice has many salutary effects: it not only induces these people to take exercise without doors, but also to eat many greens ...

[103] Edwards, *Horse and Man in Early Modern England*, 277.
[104] De Valagin, *A treatise on diet*, 214.
[105] Ibid., 84.

which they would never think of purchasing'.[106] Such a gardening scheme was, certainly at the time of Buchan's writing, not available in other cities, but Buchan and his contemporary colleagues were adamant that such a return to nature was exactly what sedentary city men needed to regain their health.

'Fortunate are they to whose lot it has fallen to remain near and true to their present Earth; and to find their happiness, labour, and destination in immediate intercourse with Nature!', Hufeland exclaimed, reminding his readers of the effect country living could have on health and happiness.[107] Despite the obvious health benefits associated with this lifestyle, physicians saw this way of life as becoming less and less practised:

> [I]t is a melancholy reflection, that this kind of life, the earliest and most natural state of man, should at present be so little esteemed that … the proportion between countrymen and citizens seems daily to be diminished. It certainly would be better for the happiness of individuals, as well as of the whole, if the greater part of the pen-knives and scissors now in use were converted into plough-shares, and if those hands occupied in scribbling were employed with the plough and the spade.[108]

'All indeed cannot be farmers by profession', acknowledged Hufeland, 'but how beneficial would it be if men of letters, people of business, and those who labour with their hands, would divide their times between both kinds of employment'.[109] Charles Scudamore, who blamed too much sitting for the resurgence of chronic disorders such as gout, pointed to cases where patients had been cured from the gout after retiring from their sedentary occupations and moving to the countryside.[110] Physicians did not necessarily recommend patients forsake all city business; however, Hufeland instead recommended that 'all the melancholy consequences of a sedentary life' would disappear if people just spent a few hours every day, or a few months in the year, labouring or exercising.[111] As we can see from this argument, the problem with sedentary employments in the city was

[106] Ibid., 57.

[107] Hufeland, *The Art of Prolonging Human Life,* 263.

[108] Ibid., 265.

[109] Ibid., 266.

[110] Charles Scudamore, *A treatise on the nature and cure of gout and rheumatism* (London: Longman, Rees, Orme and Brown, 1819), 481.

[111] Hufeland, *The Art of Prolonging Human Life,* 266.

that they did not allow for much balance between being still and moving, between being indoors and outdoors. In a time where the idea of balance was still very much part of the medical philosophy, such a one-sided lifestyle was viewed negatively. It was the disturbed balance that led to diseases such as corpulency, apoplexy, and the gout, which were all strongly associated with sedentary lifestyles. As such, the key to curing such diseases was to find balance once again. Such a balance, argued physicians, was to be found in the country lifestyle.

The sentiment that the contemporary milieu was an unnatural one could be found throughout medical literature of the late eighteenth century. Eighteenth-century people considered the countryside to be much healthier than the city, a perception which is confirmed by modern historical demography.[112] The city of London suffered from relatively high rates of mortality, which, Wear has pointed out, was exposed in the contemporary Bills of Mortality published by John Graunt as early as 1662.[113] In the eighteenth century, Graunt's demographic studies were succeeded by those of Thomas Short, who believed that the very population density of London caused ill health.[114] Health advice books of the eighteenth century presented countryside living as the norm and city living as an unnatural departure from this traditional way of living. The discussions about the health threats posed by sedentary urban living appear to be very much part and parcel of the wider concerns with environmental health, in which topics such as unclean air, corrupted soil, and foreign foods featured heavily. These aspects of eighteenth-century medical debates have been discussed extensively by historians such as Andrew Wear and Maria Pia Donato.[115] Physicians of the late eighteenth century were generally very concerned with how city air, urban water sources, and food grown in or near urban

[112] Andrew Wear, "Making sense of health and the environment in early modern England", in *Medicine in Society: Historical Essays*, ed. Andrew Wear (Cambridge: Cambridge University Press, 1992), 129.

[113] Ibid.

[114] Thomas Short, *New Observations of City, Town and Country Bills of Mortality* (London, 1750).

[115] Wear, "Making sense of health and the environment in early modern England; Andrew Wear, "Place, Health, and Disease: The Airs, Waters, Places Tradition in Early Modern England and North America," *Journal for Medieval and Early Modern Studies*, 38, 3 (Fall 2008), 443–466; Maria Pia Donato, "The afterlife of the non-naturals in early eighteenth-century Hippocratism: from the healthy individual to the healthy population," in *Conserving Health in Early Modern Europe: Bodies and environments in Italy and England,* eds. Sandra Cavallo and Tessa Storey (Manchester: Manchester University Press 2017), 158–182.

soil affected health. Although many of these concerns were linked to fears around the spread of diseases, physicians were equally worried about how the human body was affected by what they perceived to be an unnatural state of living. The perceived effects of urbanisation on exercise levels, however, has been woefully underexplored.

Conclusion

Although exercise, as one of the six non-naturals, has always been part of health advice, the eighteenth century in particular is a period of interest to any scholars of the history of exercise or the history of health advice. It was in the early eighteenth century that physicians grew increasingly interested in this topic in particular, as new medical theories focussing on the solid fibres of the body ascribed an increased importance to the role exercise played in maintaining these fibres. In a medical milieu where the state of the solid fibres determined the functioning of the very circulation of the fluids, exercise shifted from being an optional, to a vital part of anyone's health regime. In light of this more prominent role, exercise advice changed in this period. Not all exercises were equal, however: those which managed to affect the most fibres were better. Horseback riding and walking therefore became recommended over all other exercises, for these afforded the body with the minute and constant 'vibrations' that physicians were looking for in effectual exercises. Iatromechanical theory therefore had a notable effect on the exercise advice of this period, both in the importance that was ascribed to exercise, and in how contemporaries thought about the effects of exercise on the body, and how this effect could best be achieved.

The prominent role that the topic of exercise enjoyed did not wane when the iatromechanical school became less influential. Late eighteenth-century physicians continued to laud the effects of exercise on the body and on health. In fact, physicians apparently grew only more concerned with people's exercise levels from the second half of the century onwards. The term 'sedentarism' appears frequently in health guides of this period, and the application of this term over that of 'idleness' denotes a shift in physicians' focus from individual exercise habits to those observed to be more widespread in society. Lazy individuals and sedentary occupations had always existed, but late eighteenth-century physicians believed that both had become rampant in their time. As shown, there are clear indications in health guides that physicians were worried about the effects on society of what they believed to be a widespread problem caused by the changing nature of eighteenth-century society. Physicians considered the effects of sedentary lifestyles on the workforce, but also on birth rates, and on the well-being of children of sedentary parents. In other words, it appears that in the second half of the eighteenth century, insufficient exercise came to be considered a societal rather than an individual issue.

The exact reason for this shift is not entirely clear, although physicians of this period attributed it to urbanisation. The diseases often associated with sedentary lifestyles were rarely found in the countryside, physicians argued, and rural areas did not suffer from the low birth rates that cities such as London did. Life in the cities often involved occupations which were inherently less physically active than those of the countryside. Moreover, where country women were often involved with rural labour, city women were said to have far less physical tasks. Sedentary occupations would not be the problem if they were balanced by exercise taken in leisure time, but this too was waning, physicians lamented. Where people in rural areas walked or rode everywhere, cities were full of carriages to carry people wherever they needed to go. Even if a person wanted to ride or walk, the busy streets of London afforded little opportunity to do so. Although there were, of course, communal green spaces, there are some indications that these might have been disappearing in favour of buildings, leaving people with fewer spaces to take exercise.

Physicians considered such a lifestyle an unnatural state of living and many health guides were filled with a nostalgia for country living. Such expressions of longing for country living were not at all unique to discussions of exercise but could be found in relation to virtually any non-natural. This aspect of late eighteenth-century discussions about exercise should therefore be considered within the context of wider environmentalist discussions found in this period. As other historians have shown, physicians were worried about the effects of city air, city water, and city soil on the body, and it seems that worries about unnaturally low exercise levels were part of this discussion about the detrimental effects of city living. This affords a more satisfying explanation of the relatively sudden concerns with societal sedentarism than the possibility of sudden and extreme changes in the nature of work. That being said, exercise has thus far remained underexplored by historians researching late eighteenth-century concerns with the environment, as well as by historians of medicine more broadly, so the reasons behind these shifts in discussions about exercise remain difficult to gauge. While this chapter has sought to raise the issue, future research into the nature of late eighteenth-century occupations, exercise and leisure environments, and city living, is very much needed.

CHAPTER 5

'Ad ruborem, non sudorem': Sweat, Insensible Perspiration, and the Limits of Healthful Exercise

This chapter will further develop the analysis and show how eighteenth-century exercise advice was a unique product of its medical milieu. Here, I will discuss eighteenth-century rules about what constituted 'healthful', 'moderate' or 'immoderate', 'violent', or 'excessive' exercise, and place these rules in their relevant medical contexts. The chapter will focus on the eighteenth-century belief that exercise should be 'ad ruborem, non sudorem' (to redness, not to sweat), a sentiment which can be found in health guides throughout the eighteenth century. The role of sweat as a distinguishing factor separating moderate from immoderate exercise is one of several factors which can make pre-modern exercise advice seem so far removed from our own modern ideas about what exercise is where the limits of healthy exercise habits lie. As I will show, sweat and insensible perspiration played an important role in eighteenth-century exercise advice as physicians discussed it to help readers of exercise advice determine how and how much they should exercise. This in itself is good reason to investigate the eighteenth-century relationship between sweat and exercise more closely than has been done hitherto.

Some excellent work exploring the pre-modern relationship between sweat and exercise already exists. Sandra Cavallo and Tessa Storey found that in the mid-sixteenth century, Italian physicians had started to attribute greater status to 'light exercises' such as walking or even standing, as

M. Wübbels, *Quantification and Measurement in British Health Advice, c. 1670–1830*, Palgrave Studies in Medieval and Early Modern Medicine,
https://doi.org/10.1007/978-3-032-32760-4_5

part of a new medical ethos which feared sweat and exhaustion.[1] Cavallo and Storey discovered that in mid-sixteenth-century Italian texts, Galen's emphasis on strengthening disappeared from discussions, while 'frictions' (vigorously rubbing the skin with a cloth) were more likely to be recommended to altogether replace active exercise for the weak, the old, the very young, and the sick.[2] In these texts, physicians discussed the potential dangers of exercise, focussing on the belief that movement consumed vital heat, which could exhaust the body and accelerated old age. James Kennaway and Rina Knoeff recently argued that by the end of the eighteenth century, exercise advice had started to move away from this emphasis on gentle movements, towards a return to 'strengthening' exercises, once again echoing Galen's advice that only movements which were swift or vigorous truly counted as exercise.[3] The cause for this revival of the Galenic model of strengthening exercise, Kennaway and Knoeff argue, was the eighteenth-century belief that the body consists of a network of interwoven fibres, all of which need to be strengthened through exercise to keep their natural elastic properties from weakening.[4] As part of this shift, ideas of what counted as moderate or excessive exercise changed as well. Kennaway and Knoeff argue that some physicians of the late eighteenth and early nineteenth centuries no longer considered sweating to be a sign of overexertion.[5] This study finds that concerns with sweat were prominent throughout the eighteenth century. In this chapter I will demonstrate that Santorio's sixteenth-century discovery of insensible perspiration as the most considerable and the most perfect of excretions was used to reinforce eighteenth-century concerns about sweating, which by many physicians was considered an imperfect, and even 'violent' form of excretion which could disrupt the natural flow of the excretory system. These concerns found expression in discussions about exercise in eighteenth-century health guides, where visible sweat was still often presented as

[1] Sandra Cavallo and Tessa Storey, *Healthy Living in Renaissance Italy* (Oxford: Oxford University Press, 2013), 146–150.

[2] Ibid., 148.

[3] James Kennaway and Rina Knoeff, '"For it is the debilitating fibres that exercise restores": Movement, morality and moderation in eighteenth-century medical advice literature' in James Kennaway and Rina Knoeff (eds) *Lifestyle and Medicine in the Enlightenment: The six non-naturals in the long eighteenth century* (Abingdon: Routledge 2020), 108–111.

[4] Ibid., 111. For a more detailed explanation of this 'fibre body', see chapter four, especially pp. 114–115.

[5] Ibid.

something which was detrimental to health and therefore ought to be avoided.

This chapter will also assess how exercise advice fits in with wider concerns with the balance of *ingesta* and *excreta* and what this meant for readers of health literature, who were understood to all have different bodies and different levels of perspiration. Physicians did not apply the rule 'ad ruborem, non sudorem' universally or without consideration for the different perspiration levels and needs of their readers. In a similar fashion to their dietary advice, physicians generally concluded that exercise needs were different across the wide variety of constitutions, ages, and illnesses, and that some bodies in some cases did need to sweat. This chapter will argue that exercise advice, like dietary advice, retained much of its bespoke nature in that physicians continued to consider factors such as age and illness, especially when dispensing exercise advice, while other factors such as constitution started to matter less. To demonstrate this, I will look at exercise advice aimed at men, women, the old and the young, the healthy and the ill, and at different constitutions.

Finally, this chapter will consider why this particular aspect of exercise discussions survived into the late eighteenth century. Santorio's assertion that the body could be weakened beyond repair if it sweated too much originated in humouralism and can still be found in late eighteenth-century vitalism, even if the terms and explanations changed to fit the latest medical theories. Vitalism is a doctrine which, like iatromechanism, can be difficult to define due to the varied applications of the term in both the past and present.[6] As such, there is some disagreement among scholars when the term is or is not applicable and how we ought to distinguish between various kinds of vitalist theories.[7] Among historians of eighteenth-century medicine, vitalism often refers to the theory that life is generated and sustained through some form of non-mechanical power located in living bodies. As Catherine Packham explained, eighteenth-century vitalists argued that bodies possess vital energies of self-generation and the ability to take actions.[8] According to vitalists, life was not sustained by the mechanisms of an automaton, as many iatromechanics had argued, but by

[6] Charles T. Wolfe, "Vitalism and the Metaphysics of Life", in *Life and Death in Early Modern Philosophy*, ed. Susan James (Oxford: Oxford University Press, 2021), DOI: 10.1093/oso/9780192843616.003.0016

[7] Ibid.

[8] Catherine Packham, *Eighteenth-Century Vitalism: Bodies, Culture, Politics* (Basingstoke: Palgrave Macmillan, 2012), 1–2.

a repository of vital energy.[9] In the context of eighteenth-century health literature vitalists concerned themselves especially with the question of how life could be extended for as long as possible through the careful management of the non-naturals in general and diet and exercise in particular. Though, as this book shows, iatromechanical ideas about how long life could be achieved continued to be expressed, these increasingly came to be challenged and replaced by vitalist theories from the mid-eighteenth century onwards.[10] Unlike many other aspects of iatromechanism, concerns with sweat were present in vitalist health advice as well. This, I argue, explains why the fear of sweat can still be found in even late eighteenth-century texts, when other Santorian anxieties, such as that of minute shifts in weight, became much less expressed in late eighteenth-century medical literature.

Sweat and Insensible Perspiration

Santorio did not invent the concept of insensible perspiration, but his work linked the pursuit of health with the monitoring of bodily discharges—including sweat and especially insensible perspiration—and the continued popularity of his *Medicina statica* ensured that insensible perspiration remained at the forefront of seventeenth- and eighteenth-century discussions about health and health maintenance. His finding that insensible perspiration was the most significant of all the body's excretions (sweat, urine, and stool being the other and 'lesser' excretions) caused him and many physicians after him to focus their attention on insensible perspiration over all other excretions. By the mid-seventeenth century, as Lucia Dacome put it, perspiration had acquired 'a ubiquitous status in the vocabulary of health'.[11] If health depended on the due evacuation of waste matter, it stood to reason that physicians should pay attention especially to that excretion which was most considerable. There was little point to adjusting the *ingesta* to the *excreta* if the value of the latter was only partially known, after all. This shift in emphasis from all the excretions to insensible perspiration particularly, caused physicians to put insensible

[9] Ibid.

[10] Theodore M. Brown, 'From Mechanism to Vitalism in Eighteenth-Century English Physiology', *Journal of the History of Biology*, 7, 2 (Autumn 1974), 179–216.

[11] Lucia Dacome, "Balancing Act", *Studies in History and Philosophy of Biological and Biomedical Sciences*, 43, (2012), 387.

perspiration at the very centre of discussions of health. We have already seen in previous chapters the effect of this focus on insensible perspiration had on dietary advice: physicians such as Cheyne and Cocchi constructed entire dietary regimes around foods which encouraged insensible perspiration.[12] The seventeenth- and eighteenth-century focus on insensible perspiration had tremendous effects on exercise advice as well.

In seventeenth- and eighteenth-century medical literature, perspiration and sweat were often distinguished from one another, though there is some scholarly disagreement regarding the extent to which physicians differentiated between sweat and perspiration. As E.T. Renbourn noted in what is perhaps the first essay on the history of sweat, Galen differentiated between perspiration and sweat, calling the former 'innate' to health and the latter a 'preternatural' discharge.[13] Nevertheless, Renbourn argued that it was specifically from Santorio's experiments onward that, generally, the term 'perspiration' was used to mean 'not visible droplets of sweat but only the invisible vapours of the body'.[14] More recently, Michael Stolberg argued that most early modern authors believed the matter of sweat and insensible perspiration to be 'essentially the same', with insensible perspiration simply reflecting the excretion of the finer and more volatile parts of what was essentially the same matter as sweat.[15] However, many authors of late seventeenth- and eighteenth-century English medical texts explicitly differentiated between the two and described perspiration as beneficial, and sweat as detrimental, to health, which had a notable effect on exercise advice of this period.

This differentiation between sweat and insensible perspiration was expressed by Santorio himself in his 1614 *Ars de Medicina statica*, in aphorism XXI, which read: 'That Perspiration which is beneficial, and most clears the Body of superfluous matter, is not what goes off with Sweat, but that insensible Steam, or Vapour'.[16] Santorio agreed with Galen's

[12] See Chap. 2 of this book.

[13] E. T. Renbourn, "The Natural History of Insensible Perspiration: A Forgotten Doctrine of Health and Disease", *Medical History*, 4, 2 (1960), 136.

[14] Ibid., 139.

[15] Michael Stolberg, "Sweat. Learned Concepts and Popular Perceptions, 1500–1800", in *Blood, Sweat and Tears: The Changing Concepts of Physiology from Antiquity into Early Modern Europe*, eds. Manfeld Horstmanshoff, Helen King, and Claus Zittel, (2012), 506–507. https://doi.org/10.1163/9789004229204_022.

[16] Santorio Santori, in John Quincy (ed.) *Medicina Statica, Being the Aphorisms of Sanctorius, Translated into English, with Large Explanations. To which is added, Dr Keill's*

assertions that sweat was not produced by the body naturally but always came about from an external cause such as violent exercise, or a fever, and it almost always had a negative effect on the other, natural evacuations, especially insensible perspiration. Santorio stated: 'Sweat is always from some violent Cause, and as such (as Static Experiments demonstrate) it hinders the insensible Exhalation of the digested Perspirable Matter'.[17] Sweat was described by Santorio as crude matter which had not been fully refined by the digestive process, whereas insensible perspiration was the end product of the perfectly executed digestive process. Exercise, if taken too far, could prematurely force out the crude matter in the form of sweat. Santorio explained:

> That which passes the Skin by violent Exercise, is the Perspirable Matter and Sweat together; but as it is violent, it raises a great deal of undigested Juices; for it seldom happens, that so much well digested perspirable Matter is at once collected in the Body, as is the Quantity by Violence expelled.[18]

Exercise which resulted in visible sweat constituted a disruption of the digestive process, and it risked robbing the body of undigested matter which otherwise could still nourish it. Santorio therefore warned his readers that sweating diminished the strength of the body, caused the body to become lighter, hastened old age, and threatened 'untimely Death; for, according to the Philosopher, those who are exercised die sooner than those who are not'.[19] Not all exercise was bad, however. In fact, moderate exercise was understood to promote insensible perspiration and was beneficial to health. Santorio recommended 'moderate dancing, without jumping', and walking, for these exercises 'leisurely expel the digested perspirable Matter'.[20]

This differentiation between sweat and perspiration was not limited to Santorio but can be found throughout the eighteenth century. In his English translation of Santorio's work, apothecary and author John Quincy (unknown–1722) concurred with Santorio that 'it is very necessary to distinguish between Perspiration and Sweat, they differing so much

Medicina Statica Brittanica, with comparative Remarks and Explanations. As also Medico-Physical Essays. The Fifth Edition (Pater-Noster-Row: Printed for T. Longman, 1737), 54.

[17] Santorio, Medicina *Statica* (version 1737), 255.

[18] Ibid., 56.

[19] Santorio, *Medicina Statica* (version 1737), 263.

[20] Ibid., 270.

from one another, that as one is useful and preserves Health, the other is always injurious and destructive of it, unless when 'tis to give Relief from some greater Evil; as Fever, or the like'.[21] Perspiration made the body 'lightsome and cheerful', while sweat made it 'faint and dispirited'.[22] The effects of sweat and perspiration on the body differed, Quincy explained, because they were 'differently Supply'd', perspiration being the end result of 'a long Course of Circulations' and therefore being 'divested of all that can be of further Service to any Part of the Body', while sweat was 'supply'd more immediately from the Blood' and therefore contained nourishing matter still, was thicker of consistency, and 'robbed' the body 'of a great deal of Nourishment'.[23] Quincy explained that perspiration was the product of a fully and perfectly executed process of digestion. It consisted of matter that had fully been processed through all the stages of digestion and therefore contained only very fine waste matter.[24] Other forms of excretions (sweat, urine, or stool) took place at earlier stages of digestion and could therefore still contain nourishing matter—sometimes rendered noxious through corrupted or interrupted digestion.

Quincy explained that the potential ill effects of sweating on the body were twofold. First, the body became weakened due to the loss of potentially nourishing matter that the violent evacuation through sweat had caused. Second, the weakening of the body consequently impacted its very ability to perspire insensibly. Extrapolating from Santorio's research, Quincy elaborated: 'the more a Person Sweats, it is certain that so much less he perspires', because the body's ability to perspire depended on the strength and health of the body's vessels.[25] A strong and healthful body possessed the ability to digest its nourishment perfectly and effectively enough to ensure food was turned into very fine waste matter which would be cast off insensibly through the pores of the skin as insensible perspiration. But if the body was weakened as a result of sweat depriving the body of nourishment, the digestive organs would weaken, digestion would become increasingly impaired, and the matter to be digested would be digested imperfectly and cast out while in a crude state still, or worse yet, linger in the body, causing inflammation and blockages. Insensible

[21] Quincy, *Medicina Statica* (version 1737), 54–55.
[22] Ibid., 10.
[23] Ibid., 10.
[24] Quincy, *Medicina Statica* (version 1737), 55.
[25] Quincy, *Medicina Statica* (Version 1737), 55.

perspiration was 'a Token of Health', but sweat, Quincy argued in his own aphorisms, 'the contrary'.[26]

Quincy's contemporaries were of similar mind regarding the distinction between sweat and perspiration. Whatever was excreted from the skin besides perspiration was 'preternatural', physician Edward Strother (unknown–1737) argued, and was not part of the body's natural digestive process but had some other cause 'within us, or without us' that caused it.[27] Eighteenth-century physicians also concurred with Santorio that sweat disrupted the natural flow of perspiration and was therefore not only undesirable, but even dangerous to health. Bath physician George Cheyne (1672–1743) compared sweating to diarrhoea to make the point that sweating was an excessive and preternatural evacuation. Perspiration and sweat differ 'as widely as the daily natural emptying of our bowels and a looseness of diarrhoea', he explained. The excessive flow diminished the strength of the vessels which contained them. Diarrhoea reduced the strength of the bowels, and similarly sweating would reduce the strength of the vessels and the pores.[28] Bernard Lynch argued similarly that sweat 'differs very much from insensible perspiration; for it to be increased … Perspiration must unavoidably be diminished'.[29] Even if, as Stolberg argues, physicians believed sweat and perspiration to be formed of essentially the same matter, many physicians clearly distinguished between then, praised the effects of one and vilified the other, and believed the evacuation of sweat to be antithetical to the evacuation of insensible perspiration.

Sweat and Exercise Advice

The combination of these beliefs about the distinction between sweat and perspiration, and the iatromechanical importance placed on exercise as the primary method for keeping the fibres in due strength, resulted in a seemingly complex situation which required readers of seventeenth- and eighteenth-century health guides to carefully balance between sufficient and excessive exercise. Although the exact point at which this balance was

[26] Quincy, *Medicina Statica* (version 1737), 109.

[27] Edward Strother, *The family companion for health, or, plain, easy and certain rules* (London: Printed for F. Fayram, 1719), 408, Eighteenth Century Collections Online.

[28] Cheyne, *An Essay of health and long life*, 196.

[29] Lynch, *A guide to health through the various stages of life*, 308.

achieved differed from person to person, most people would find it by monitoring their skin closely for signs of visible sweat. Unless there was an acute need for a critical discharge, such as in the case of a fever, insensible perspiration was never to be raised to a visible sweat, Thomas Byfield warned in 1685, and he recommended that '[s]o much exercise as promotes a full perspiration, without sweating is a convenient mean, that common saying, *ad ruborem non sudorem*, is a very good rule'.[30] George Cheyne warned in 1742 that exercise should be 'not to sweating, but to warmth' and James Mackenzie recommended in 1752 that 'when any person begins to sweat, or grow weary, or short breathed, he should forbear for a while, in order to recover himself'.[31] Still in 1780, John Trusler cautioned that 'exercise be not too violent, nor continued longer than it is a pleasure; but that we desist before we be weary, or sweat too much'.[32] This idea, though popularised by Santorio and the iatromechanical physicians revisiting Santorio's work, had been around much longer though, as we can see from early seventeenth-century physician Robert Burton (1577–1640), who cited Galen:

> What kind of exercise he should use, Galen tells us ... and in what measure, "till the body be ready to sweat," and roused up; *ad ruborem*, some say, *non ad sudorem*, lest it should dry the body too much ... Some prescribe frequent and violent labour and exercises, as sawing every day so long together, but that is in some cases, to some peculiar men; the most forbid, and by no means will have it go farther than a beginning sweat, as being perilous if it exceed.[33]

[30] T. Byfield, *Two discourses one of consumptions, with their cure by a new method, the other contains some rules of health* (London: Printed by Dorman Newman, 1685), 23, Early English Books Online.

[31] George Cheyne, *The natural method of cureing the diseases of the body, and the disorders of the mind depending on the body* (London: Printed for George Strahan, 1742), 258, Eighteenth Century Collections Online; James Mackenzie, *The history of health, and the Art of preserving it* (Edinburgh: Printed and sold by William Gordon, 1758), 379, Eighteenth Century Collections Online.

[32] John Trusler, *An easy way to prolong life, by a little attention to what we eat and drink* (London: Printed for the author, and sold by J. Bell, 1775), 89, Eighteenth Century Collections Online.

[33] Democritus Junior, "Exercise rectified of Body and Mind", *The Anatomy of Melancholy* (London: Printed & to be sold by Hen. Crips & Lodo Lloyd, 1652), Project Gutenberg Online Version, The British Library Online: http://www.gutenberg.org/files/10800/10800-h/10800-h.htm.

This method for determining the limits of healthful exercise survived into the late eighteenth century, though, as Kennaway and Knoeff recently argued, the fear of sweating had started to abate somewhat around this period.[34]

Most seventeenth- and eighteenth-century health guides advised that any exercise was permissible so long as it was not excessive or too violent. However, physicians did tend to characterise only a few exercises as meeting the requirement of both sufficiently exercising the whole body and being moderate enough so as to not raise a sweat. As demonstrated in the previous chapter, the exercises which physicians recommended to their readers tended to be those which supplied the whole body (rather than just a section of the body) with constant and minute vibratory motions (rather than large, intermittent, and high-impact movements).[35] Exercises which physicians believed to achieve that were horseback riding especially, but walking or using an exercise chair was also serviceable. According to Cheyne, healthful exercise was that which was constant, uniform, and regular, not violent, but orderly, and done not to a sweat, but merely until the body was warm. Walking fit these requirements best he argued, calling it the most natural and healthy of all exercises. He explained that walking was the 'least laborious' of all exercises and, as such, was least likely to weaken the body.[36] Riding was generally deemed especially appropriate for those readers who were not in a healthful state, because it was an activity in which the body was moved by an external force, rather than moving itself. The benefit of this activity was that it was not so strenuous as to promote sweat, but the body still received those minute vibrations mechanists believed to be so imperative to the strength of the fibres.[37]

The quantification of the body's excretions had a noticeable effect on written exercise advice, and physicians of the late seventeenth and eighteenth centuries were increasingly willing to quantify how much and how often they believed persons should exercise. Though the commonly used adage that 'one rule cannot apply to all' was upheld in exercise advice, the physicians who did quantify how much exercise was healthful for most argued that exercise should be had daily. Physicians including John Burton

[34] Kennaway and Knoeff, "For it is the debilitating fibres that exercise restores", 111–138.

[35] See Chap. 3.

[36] Cheyne, *An Essay of health and long life*, 94–95, 172 and 258.

[37] Francis Fuller, *Medicina Gymnastica: or, a treatise concerning the power of exercise* (London: Printed by John Matthews, 1705), 150, Eighteenth Century Collections Online.

(1710–1771), George Cheyne, and William Buchan (1729–1805) recommended that most persons should spend two to three hours a day in exercise, adjusting the kind of exercise (walking instead of riding, for example), rather than the amount spent in it, to account for differences in physical strength and abilities.[38]

These frequencies and timeframes were not explicitly linked to Santorio's (or any other) static experiment, so it is unclear whether these quantities were based on any result of experiments measuring the effects of exercise on the excretions of the body. Santorio had demonstrated that exercise, by increasing the quantity excreted, caused the body to lose weight. In his aphorisms, he stated: 'By Exercise Bodies are rendered lighter; for all the parts, especially the Ligaments and Muscles, are cleared of their Excrements by Motion: the perspirable Matter is fitted for Exhalation and the Spirits are rendered finer'.[39] In his comments on Santorio's aphorisms, Quincy elaborated that when the body is in motion, the 'additional Force of the circulating Fluid' on the excretory organs would cause these organs to discharge more than they would when the body was in a state of rest.[40] Too much exercise would 'waste as much as is daily supplied, but also wear away a great deal of the substance of the Body', Quincy warned, though he added that such an increase in excretions could be 'of great Service when there is an over Weight, and need of Discharges'.[41] Santorio had recommended that exercise was 'most wholesome' after the first and second 'concoction' (a term used to describe food being broken down in the stomach) twice a day, before eating an amount of food which would put the body back at its 'wonted Standard', meaning, the usual daily weight of the body (before it is increased by food or

[38] John Burton, *A treatise on the non-naturals, in which the great influence they have on human bodies is set forth, and mechanically accounted for* (London: Printed by A. Staples, 1738), 257–259, Eighteenth Century Collections Online; Cheyne, *Essay on health and long life*, 95–99; William Buchan, *Domestic medicine; or, a treatise on the prevention and cure of diseases by regimen and simple medicines.* 2nd edition (London: Printed for W. Strahan, 1772), 90–92, Eighteenth Century Collections Online.

[39] Santorio, *Medicina Statica* (version 1737), 257.

[40] Quincy, *Medicina Statica* (version 1737), 258.

[41] Ibid. Here, the use of the term 'over Weight' almost certainly means a higher weight than is average for a particular individual within a 24-hour period, rather than a weight which is higher than what physicians recommended people weigh, or higher than what people weigh 'on average'. The modern concept of 'overweight' did not yet exist in this period, nor did any concept of an average weight for men or women. For further explanation, see Chap. 6.

decreased by various excretions).[42] Quincy, in his commentary on Santorio's aphorisms, considered it 'a very good Rule' but added that this could not be exactly known without the 'Balance' (meaning, a device to weigh the body) and recommended instead that people go by the *perception* of their own weight as guidance as to when they should exercise to bring their weight back to what was their standard weight in a 24-hour period.[43] Though the connection between body weight and exercise had been proven by Santorio, nevertheless body weight was not used outside of static experiments as a method for determining how much, or for how long, a person should exercise.

That being said, there are at least two examples of instances where people weighed themselves after exercise to determine how much they had excreted as a result of exercise. The first comes from the diary of Samuel Pepys, who recorded in 1667 that he had witnessed the then king Charles II weighing himself after a game of tennis. Pepys was told that this was only the King's 'curiosity', to see how much weight he lost by playing: 'and this day he lost 4.½ lbs'.[44] The second instance can be found in a letter claimed to have been written by a 'valetudinarian' to essayist Joseph Addison, who then published the letter in *The Spectator* in 1711. In it, the valetudinarian (a term then used to describe people who were generally sickly) shared his anxieties relating to balancing his *ingesta* and *excreta* and his methods for monitoring shifts between the two. After a full meal, he explains, he walks until he has perspired 'five Ounces and four Scruples', monitoring his success by means of a weighing chair.[45] These examples suggest that people were aware of the possibilities of quantifying the effects of exercise by measuring how much weight the body lost through perspiration during exercise, but whether this was practised is another matter. Pepys specifically states that the king weighed himself out of mere curiosity, which implies that it was not done for the purpose of health maintenance. The letter from *The Spectator* was intended as satire and was almost certainly written by Addison himself and therefore does not

[42] Santorio, *Medicina Statica* (version 1737), 265,

[43] Quincy, *Medicina Statica* (version 1737), 265. Emphasis my own.

[44] Samuel Pepys, 2 September 1667, in *The Diary of Samuel Pepys*, Vol. 8: 1667. Edited by Robert Latham and William Matthews. London: HarperCollins, 2000, p. 419.

[45] This letter was published by Joseph Addison in *The Spectator*, a daily publication founded by Addison and Richard Steele, Vol. 1, No. 25 (March 29, 1711), 93. Available online at: <https://archive.org/details/in.ernet.dli.2015.39152/page/n101>. Accessed 19 January 2020.

constitute an actual instance of weight measuring in relation to exercise. It could imply, however, that this is behaviour that Addison witnessed in others and consequently ridiculed. Regardless, there are very few surviving examples of individuals weighing themselves after exercise. We can therefore only assume that weighing was not yet a common method for determining how much the body perspired as a result of exercise, nor used to determine how much exercise a person needs.

Exercising Different Bodies

Although the rule 'ad Ruborem, non Sudorem' was common in eighteenth-century exercise advice, physicians did not apply it to everyone equally. In fact, some physicians made exceptions to this rule by prescribing sweat-inducing labour or exercises in some cases despite the known dangers that sweating posed to health. There were several reasons why exercise rules could not possibly be the same for all bodies: firstly, people possessed different levels of strength, whether due to age, gender, or physical (dis)abilities. Secondly, not everyone had the same opportunities for exercising: it often required time, space, and disposable income to afford equipment. Finally, people had different exercise needs depending on the balance between their *ingesta* and *excreta*. For those who ate much and were largely sedentary, exercise was considered even more necessary than for those who ate but little or who already perspired much through labour.

Exercise was a manner through which people could increase their excretions through perspiration and was therefore often recommended, especially to people who needed to excrete more. This was the case both before and during the eighteenth century, though there do seem to be some differences in which bodies were discussed in this context. In seventeenth-century texts, bodies with certain constitutions, such as those which were hot and choleric, were often cautioned against using too much exercise, while phlegmatic bodies were believed to have more need of it, as they had a predisposition to accumulating excess fluids.[46] '[T]hose bodies that are full and gross, and bodies repleted with humours, must use such exercise, as may correct the distemper of the body', one physician argued in 1674, and another similarly recommended that exercise was excellent for people

[46] Byfield, *Two Discourses.* 25; Maynwaring, *methods and means of enjoying health*, 140.

whose humours were 'viscious and gross'.[47] It was argued that exercise, by raising the heat of the body, thinned the humours and helped the body excrete these.[48] There was therefore a clear distinction between exercise advice for bodies which were of a phlegmatic constitution, which were thought to abound in moisture, and for bodies whose constitution implied dryness. Over the course of the eighteenth century, these discussions somewhat shifted away from a focus on humours to a focus on illnesses instead. Although matters of constitution did not disappear from exercise advice, physicians dedicated more space to considering how exercise affected illnesses and for which ailments exercise was appropriate or inappropriate. Illnesses which mirrored the issues associated with certain constitutions received exercise advice similar to what people of those constitutions would have been told, albeit with slightly different terms or explanations. For example, fat bodies, which in the eighteenth century were described as 'gross' and 'fill'd with blood or oil', were, like phlegmatic bodies in seventeenth-century treatises, considered to be in need of exercise especially.

It appears that fat bodies constituted an exception to the no-sweating rule in early eighteenth-century health advice. A number of physicians from this period can be found to recommend sweating to corpulent readers. For example, Jeremiah Wainewright argued in 1708 that gentlemen who indulged in eating excessively could use no better antidote against the effects of overeating than 'violent exercise'.[49] Cheyne seems to have allowed sweating for the corpulent, as can be seen from his letter to his corpulent friend Samuel Richardson (1738), who was told by Cheyne that he could ride or work at a pump 'ad sudorem'.[50] Bernard Lynch similarly argued that 'they that are lean should exercise till the body and spirits are gently heated ... but they who are fat may exercise ad sudorem, that is, till they sweat'.[51] Corpulency as an illness had been rarely discussed in medical

[47] A.B., *The sick-mans rare jewel wherein is discovered a speedy way how every man may recover lost health, and prolong life* (London: Printed by T.R. and N.T, 1674), 27. Early English Books Online. 27; Strother, *An essay on sickness and health*, 41.

[48] A.B., *The sick-mans rare jewel* 27; Strother, *An essay on sickness and health*, 41.

[49] Wainewright, *A mechanical account of the non-naturals*, 156.

[50] George Cheyne to Samuel Richardson, 1738, *The Letters of Doctor George Cheyne to Samuel Richardson (1733–1743), ed.* Charles F. Mullett (Columbia: University of Missouri, 1943), 42.

[51] Byfield, *Directions tending to health and long life*, 15; Lynch, *A guide to health through the various stages of life*. 290.

treatises until it gained more attention in the eighteenth century as physicians such as Thomas Short (in 1727) and Malcolm Flemyng (in 1760) started to explore its possible causes and cures. Corpulency could run in families, but, Flemyng believed, was much oftener acquired by a manner of living, such as one consisting of a rich diet and sedentary life.[52] Short described fatness, or corpulency, as 'a great Plenty of Blood, stor'd with oily Parts, and not sufficiently attenuated and discharged by Perspiration' and similarly believed this to be primarily caused by excessive eating.[53] In other words, fat bodies were argued to be full with excess *ingesta*, which needed to be excreted.

Exercise had long been considered the natural counterbalance to diet or *ingesta* due to its ability to help digest and cast out what a person had put on by eating. Strother explained:

> If you eat much, you must Work, or Exercise much; and if you use much Exercise, you must eat heartily. These necessarily infer one another; Exercise wastes us, and Victuals repair us...and if we would live in Health, we must pay the greatest Regard to these: Eating without Exercise is Cramming...for we must fall a Sacrifice to Excess, and die with Fat: As Exercise then without Victuals will consume you, Victuals without Motion will fill your Veins with gross Humours.[54]

If the natural evacuations of the body did not manage to excrete this excess sufficiently, either due to the sheer volume of it or because of some physiological defect such as naturally weak solids, the excess had to be evacuated through other means. Stools were a natural way to excrete of course, and purgatives occasioning diarrhoea would offer a speedy and effective excretion of excess, Flemyng considered, but this would weaken the solids through depriving the body of its nutrients and hurting the stomach and bowels.[55] Sweat however, Flemyng argued, contained even more waste matter than the other excretions, and promoting it could be

[52] Malcolm Flemyng, *A discourse on the nature, causes and cure of corpulency* (London: Printed for L. Davis and C. Reymers, 1760), 6, Eighteenth Century Collections Online.

[53] Thomas Short, *A discourse concerning the causes and effects of corpulency: together with the methods for its prevention and cure* (London: Printed for J. Roberts, 1727), 10, Eighteenth Century Collections Online.

[54] Strother, *An essay on sickness and health*, 46.

[55] Flemyng, *A discourse on the nature, causes and cure of corpulency*, 16.

done relatively safely through exercise.[56] For this reason, he considered exercising to be one of the most effective ways to reduce corpulency, second only to dietary moderation.

Other physicians however continued to emphasise the dangers involved with sweating even in these discussions of fat bodies. Echoing Santorio's argument that imperfect evacuations such as sweat weakened the body and its ability to perspire normally, Thomas Short stated in his treatise, *A Discourse Concerning the Cause and Effect of Corpulency* (1727): 'There must be no constant or frequent excessive Evacuations, for these hasten the Fluids out of The Body, before the Secretions be finished, or the nutritious Parts be applied and assimilated to the Sides of the Vessels for their Nourishment or Growth'.[57] Physicians had to choose between the lesser of two evils, and whether a patient ought to sweat or not depended on the degree of their fatness. As will be discussed in the next chapter, the point at which a fat body was also an unhealthy body was ill-defined for most of the eighteenth century and involved no quantifiable measures such as weight, clothing size, or waist size. Instead, physicians largely relied on gauging a patient's ability to perform their day-to-day duties and partake in moderate exercises without pain or discomfort. If these duties and activities were significantly impacted by their fatness, the patient was judged to be in a 'morbid' state of corpulency and in need of medical intervention. It was not until patients had reached this level of corpulence that physicians believed the possible benefit of a quick evacuation of *excreta* outweighed the risk posed by sweating.[58] This risk/benefit approach was not exclusive to corpulency: John Quincy also suggested that a sweat might be appropriate in the cases of acute fevers, for example.[59] Sweating was therefore not considered to be necessarily healthy for fat bodies, but it was occasionally prescribed as a method of curing an illness which posed a greater threat to health than sweating itself.

Matters of gender did come up in eighteenth-century exercise advice; however, sex difference in itself seems to have played as little a role in exercise advice as it did in dietary advice. In older, Galenic texts, it was sometimes suggested that women needed more exercise than men on account of their humoural differences, but this notion does not seem to

[56] Ibid., 17.
[57] Short, *A discourse concerning the causes and effects of corpulency*, 20.
[58] Trusler, *An easy way to prolong life*, 88.
[59] Quincy, *Medicina Statica* (version 1737), 54–55.

have survived into the eighteenth-century health guides. According to early modern medical theory, heat played a primary role in managing the circulation and expulsion of the body's fluids, including insensible perspiration. Women were said to grow fat more easily because their bodies were generally colder and moister than men's, Michael Stolberg explains.[60] Sarah Toulalan has similarly shown that women's colder constitutions were thought to contribute to their greater fatness relative to men, as early modern physicians explained that cold renders fat solid.[61] Plethora (or excess moisture) was not gender-exclusive but gender assumptions do seem to have played some role in exercise advice of this period. For example, Girolamo Mercuriale's *De arte gymnastica* (1569) stated that those who have a predominantly moist temperament particularly benefit from exercise: 'Hence among women, who are generally endowed with a moist temperament, it is those who work and exercise longer and more vigorously who live a healthier and less troubled life'.[62]

Women were told to exercise no less, but not necessarily more than men, despite this perception that they had a natural surplus of moisture. While there has not been much research done on exercise advice and practices for women of this period, a rich scholarship exists on early modern medical perceptions of the female body. As this scholarship has shown, early modern physicians believed that in women's bodies the surplus of moisture served a particular function, namely, to nourish the foetus in utero, and after birth, with breastmilk.[63] When a woman was not pregnant or breastfeeding, this surplus would collect in the womb, to be expelled monthly in the form of menstrual fluid.[64] Regular menstrual cycles were therefore seen as nature's way of expelling the superfluous moisture/humours inherent in women.[65] Women's surplus of moisture/humours was thus not considered to be as threatening to health as the plethora in

[60] Michael Stolberg, "'Abhorreas pinguedinem': Fat and obesity in early modern medicine (c. 1500–1750)," *Studies in History and Philosophy of Biological and Biomedical Sciences*, 43 (2012), 376.

[61] Sarah Toulalan, "'To[o] much eating stifles the child': fat bodies and reproductions in early modern England", *Historical Research*, 87, 235 (2013), 73–74.

[62] Citation taken from Von Malinckrodt and Schattner, *Sports and Physical Exercise in Early Modern Culture*, 150.

[63] Wendy D. Churchill, "The Medical Practice of the Sexed Body: Women, Men, and Disease in Britain, circa 1600–1740", *Social History of Medicine*, 18, 1 (2005), 12

[64] Ibid.

[65] Ibid.

corpulent bodies. While the female body might have been less able to maintain a healthy humoural balance than a male body, it was able to set right this imbalance through its own physiology.[66] The difference between the plethoric female body and the plethoric corpulent body was that the latter was considered a state of illness, while the former was not. As the female body was not in need of correcting, it was not in need of sweating either—unless the female body was also a morbidly corpulent body.

This form of expelling excess moisture was not exclusive to women. Gianna Pomata noticed how periodic bleeding in men, often taking the form of haemorrhoids, was compared to and paralleled with women's menstrual bleeding.[67] In these cases, periodic bleeding in men was considered a positive phenomenon, an indication of the male body adopting a process found in the female body to cleanse itself of any humoural excess and restore the body's balance.[68] The Spanish physician Andres a Laguna wrote in 1551 that the men who experienced such 'purgations' were 'effeminate on account of diet and lack of exercise'.[69] Moderate exercise stimulated insensible perspiration and therefore helped the body evacuate excesses, and so a total lack of exercise was a possible cause of a plethora. If the body could not expel naturally it would find alternative routes—in this case through a 'hemorrhoidal flux'.[70] Women were naturally more moist, but not considered 'plethoric' until their natural ways to evacuate this moisture through menstruation or breastfeeding ceased, whether naturally, through menopause, or through illness.[71] Swiss physician Samuel-Auguste Tissot (1728–1797) advised that 'when evacuations naturally cease' women's bleeding ought to be redirected, their food intake decreased, and their exercise increased, in an effort to stop the build-up of

[66] Gianna Pomata, "Menstruating Men: Similarity and Difference of the Sexes in Early Modern Medicine", in *Generation and Degeneration: Tropes of Reproduction in Literature and History from Antiquity through Early Modern Europe*, eds. Valeria Finucci, and Kevin Brownlee (Durham, NC: Duke University Press, 2001), 116. https://doi.org/10.1215/9780822380276-005.

[67] Ibid., 124.

[68] Ibid., 113–116.

[69] Ibid., 113.

[70] Ibid.

[71] Samuel Auguste Tissot, *Advise to the people in general, with regard to their health ... Translated from the French edition of Dr. Tissot's Avis au people &c. Printed at Lyons, with all the notes in the former English editions, and a few additional ones, by J. Kirkpatrick* (Philadelphia: Printed by John Dunlap, 1771), 364, Eighteenth Century Collections Online.

menstrual blood.[72] Scottish physician William Cullen (1710–1790) also regularly prescribed exercise in cases of menstrual obstruction, often in combination with medication that would help expel excess fluids from the body, such as diuretics, laxatives, and emetics.[73] Women were given similar advice for other obstructions that prevented timely evacuation of bodily fluids. The focus of such advice was not a matter of gender, but rather on the concept of humoural obstructions. Men who suffered from other forms of obstruction were also often prescribed exercise. Gender physiology thus seems to have played a relatively minor role in eighteenth-century exercise theory, which explains why we rarely see it mentioned in the health guides. Occasionally physicians did suggest that it was women especially who were in need of exercise, but this was argued to be due to their lifestyles rather than their constitutions, as we have seen in the previous chapter.

The advice to avoid sweat seems to have applied to women too, even if the authors of advice manuals rarely addressed women. Using the male body as a default was standard practice in medical works of the period and does not mean that the advice given in these guides was meant for male bodies exclusively. Physicians would generally discuss the female body only where their sex was deemed relevant. Pregnancy, menopause, and ailments unique to the female body were topics in which we find exercise advice which directly discussed women's bodies. It appears that women were expected to follow the advice in general health guides unless they needed advice related specifically to sex-specific health issues such as pregnancy. From letters exchanged between Cullen and his female patients we can generally glean that women were recommended the same exercises as men. There do appear to be more occasions of Cullen prescribing passive exercises such as riding in a chariot to women than there are occasions of him prescribing the same to his male patients, so it could be the case that passive exercises were considered more suited to women than to men. This may have been the result of old humoural notions that the woman's body was 'weaker' and therefore more suited to lower-impact exercises. However, in his letters, Cullen makes no mention of prescribing passive exercises on account of his patients being women. Cullen may have advised

[72] Ibid.

[73] For some examples, see the following Doc ID's 1588; 4071; 3727; 4695; 4520; 827; 5192; 2448, from "The Consultation letters of William Cullen", Royal College of Physicians at Edinburg. Accessible online via *The Cullen Project*, <http://www.cullenproject.ac.uk.

passive exercises on account of these women suffering from health conditions, rather than on account of their sex. Alessandro Arcangeli points out that passive exercises were prescribed especially to women in the later stages of pregnancy, out of fear that more active motions might cause miscarriage.[74] Regardless, passive exercises were not intended to be the norm for all women: women too were prescribed active exercises such as riding, and riding practices among women in the eighteenth century have been well documented.[75]

Age did feature often in exercise advice of this period, just as it did in dietary advice. In traditional medicine, age had always been described as a vitally important consideration when making any sort of health recommendations, and exercise was no exception. The bodies of the young and the old were understood to have unique humoural characteristics, and even as discussions of humours slowly waned in the eighteenth century, the idea that children, the elderly, and those in the prime of their lives required advice adjusted to their age groups remained. Children did not usually feature in health guides of the eighteenth century as these guides were generally written with the adult body in mind. Advice specific for children often featured in treatises which explicitly focussed on children's physick. It is possible that children were perceived as less in need of exercise advice due to their getting exercise through play. This did change somewhat towards the end of the eighteenth century, as physicians seemed increasingly worried about the long hours children spent seated behind their classroom desks.[76] Old people, however, did feature in these general health guides, which perhaps is reflective of the audience that physicians writing these guides were expecting or hoping to attract.

The health guides of the eighteenth century nearly always recommended people in old age to remain physically active, but to exercise cautiously and moderately. Cheyne argued that 'a man past or near his fifties should exercise the same as he did in his youth', continuing to ride

[74] Alessandro Arcangeli, "Exercise for Women", in *Sports and Physical Exercise in Early Modern Culture*, eds. Rebekka Von Mallinckrodt and Angela Schattner (Abingdon: Routledge 2016), 151–153.

[75] Kirsten Olsen, *Daily Life in 18th-Century England* (Westport, Greenwood Press, 2017), 148; Cally Blackman, "Walking Amazons: The Development of the Riding Habit in England during the Eighteenth Century", *Costume*, 35, 1 ((2001), 47–58. https://doi.org/10.1179/cos.2001.35.1.47.

[76] See Chap. 4 of this book for a more extensive discussion about concerns with sedentarism in health advice.

horseback, 'whilst his strength is equal to it'.[77] If he was not strong enough to ride, exercise was to be had by means of a chariot, gentle walks, or even a chamber horse—but 'some exercise, in some shape or other, must be had'.[78] The exercises recommended to older people were generally of a more passive nature. Riding in a carriage, as well as rubbing with oil or ointments, was often recommended in eighteenth-century health guides.[79] Horseback riding was fine as long as the patient could mount the horse. Due to its stimulating effects on the fibres of the body, it was sometimes considered a remedy against ageing.[80]

Of course, almost all healthful people were advised to exercise moderately, but for older readers this was even more important than for younger ones. According to humoural theory, the heat and moisture in the body steadily declined over the course of a lifetime until they were used up completely, a process often likened to the burning-up of a candle or oil lamp. Following that logic, older bodies contained less of that vital fluid that kept the lamp burning than younger bodies. Furthermore, their solids were dryer and more brittle and became increasingly more difficult to maintain and restore. Within this context, it is unsurprising that physicians advised caution when it came to exercising old bodies. The metaphor of the lamp lost some of its popularity in the eighteenth century although it was still to be found, and mechanists such as Quincy turned to comparing the ageing process instead to the wearing out of the parts of machinery:

> Old Persons die because their Solids are quite worn out, that is, they have so far lost their Texture to admit any further Supplies of such Juices as are necessary to keep them in Motion. As when the principal Wheels of a Clock are worn out, and they are capable no longer either of moving or being moved themselves, 'tis necessary the whole Machine must stand still.[81]

Old bodies were not as able to circulate and draw from the nourishing juices as young bodies and were therefore much weaker. Moderate exercise could help strengthen the weakening solids, but sweat-inducing

[77] Cheyne, *The natural method*, 89.

[78] Ibid., 89 and 301.

[79] Christopher Hufeland, *The Art of Prolonging Human Life* (London: Published by Simkin and Marshall, Stationer's Court, 1829), 325.

[80] Thomas Jameson, *Essays on the changes of the human body at its different ages* (London: Printed for Longman, Hurst, Rees, Orme, and Browne, 1811), 355–356, Archive.org.

[81] Quincy, *Medicina Statica* (version 1737), 92.

exercise only weakened them further and at a quicker rate. Santorio was quite explicit in linking excessive exercise to ageing quickly. Violent exercise 'hastens old Age and untimely Death' he said, for, '[b]y too much exercise the Fibres become hard; whence Old Age, which is an universal Hardness of the Fibres [ensues]'.[82] In the eighteenth-century translation of *Medicina statica*, Quincy elaborated:

> Too much exercise will break the nervous Fluid, so much as to occasion a greater Waste of it than can be supplied; and therefore the ... Fibres, for want of this animal Oil, will lose their Power of Motion and stand still, which is Death; and the nearer a Person is brought by any means whatsoever to this State, the further such a Person be said to be advanced towards Old-Age.[83]

The rule of exercising *ad ruborem non sudorem* was therefore even more pertinent to readers past the age of 50 than it was to most other groups.

Wasting Away

This fear around wasting matter was not new or exclusive to Santorio or to the iatromechanical school but developed out of Galenic ideas. What was new was Santorio's theory that insensible perspiration was the greatest of all excretions. In fact, Santorio had argued that 'insensible perspiration alone discharged much more than all the other excretions put together'.[84] Though not all eighteenth-century physicians agreed with Santorio's measurements of the various *excreta*, his claim that insensible perspiration was one of the most, if not the most, considerable of the *excreta* seems to have been generally accepted in the early eighteenth century. In his preface to his translation of the *Medicina statica*, Quincy argued that Santorio, 'by the Balance', had done enough to convince 'any serious Person' of the proportions of the natural Discharges, 'the most considerable of which', was that by insensible Perspiration, and from the health literature contemporary to him it appears that Quincy's claim was not unfounded.[85] For example, Jeremiah Wainewright wrote: 'I shall observe from Sanctorius, [t]hat insensible Perspiration is double to all the sensible Evacuations

[82] Santorio, *Medicina Statica* (version 1737), 263 and 270.
[83] Quincy, *Medicina Statica* (version 1737), 270.
[84] Santorio, Aphorism IV, *Medicina Statica* (1737 version), 44.
[85] Quincy, *Medicina Statica* (1737 version), vi.

made by Urine and Stool put together ... therefore a Man will be made no weaker by having Ten Times as many Stools as he us'd to have, than he will by perspiring only double the quantity he does at other times'.[86] As far as Cheyne was concerned, Santorio's experiment had ascertained that the quantity of this particular excretion was 'at least equal' to both urine and stool, which Cheyne subsequently used to support his argument as to the importance of maintaining 'free and full flowing' insensible perspiration.[87] In his re-trial of Santorio's experiment, James Keill (1673–1719) found that, in the Scottish climate, the quantity excreted through perspiration was smaller than Santorio had measured in the Italian climate and concluded that 'the most considerable Outlets, are the Anus, Kidneys, and the Pores of the Skin'.[88] Which of these three forms of *excreta* was the most considerable depended on a number of factors relating to environment and lifestyle, but temperature and exercise affected perspiration levels especially, Keill concluded similarly to Santorio.

The effects of Santorio's treatise on insensible perspiration on subsequent exercise advice were therefore considerable and understanding how the doctrine of insensible perspiration intersected with exercise advice is paramount to understanding seventeenth- and eighteenth-century exercise practices. These health guides and the advice within them are indicative of an exercise culture which revolved around carefully managing sweat and insensible perspiration for optimum health. The fear of blocked evacuations has been well-documented by historians of early modern medicine. Shigehisa Kuriyama, for example, has demonstrated that fear of excess humours lingering and putrefying in the body drove physicians to argue that insufficient insensible perspiration was detrimental to health, and we see this fear clearly reflected in the health guides of the eighteenth century.[89] For example, Scottish physician William Buchan stated in his *Domestic Medicine* (1769): 'Insensible perspiration ... is of so great importance to health, that few diseases attack us while it goes properly on; but when it is obstructed, the whole frame is soon disordered'.[90]

[86] Jeremiah Wainewright, *A Mechanical account of the non-naturals* (London, printed for Ralph Smith, 1708), 145–146.

[87] Cheyne, *Essay of Health and long life*, 128.

[88] Keill, *Medicina Statica Britannica*, 'Dedication', and 242–247.

[89] Shigehisa Kuriyama, "The Forgotten Fear of Excrement", *Journal of Medieval and Early Modern Studies*, 38, 3 (Fall 2008), 413–442. DOI 10.1215/10829636-2008-002.

[90] Buchan, cited in Kuriyama, "The Forgotten Fear of Excrement", 417.

Obstructions were not the only danger associated with insensible perspiration, however. Just as dangerous were disruptions caused by excessive evacuations from the other *excreta*, which, as we have seen, included sweating. According to Santorio, frequent violent exercise could cause the body to be deprived of the nutritive matter that it needed to grow, maintain, and restore the solids of the body, thus eventually leading to a 'wasting' of the solids of the body, which, in turn, could lead to premature ageing and even death.[91] This idea found expression in a number of eighteenth-century health guides, despite the general eighteenth-century consensus that exercise was beneficial to health. In his eighteenth-century translation of Santorio's *Medicina statica*, Quincy added his own thoughts on the matter: 'A continuance of Exercise beyond the Strength of the Constitution, will not only waste as much as is daily supplied, but also wear away a great deal of Substance of the Body, and thereby sink both its Strength and Bulk'.[92] Thomas Byfield also argued that overly exerting exercise would 'waste the flesh, fire out the body's natural heat, and disrupt the due fermentation of the blood'.[93] It was thus not only the nature of the exercise but also its quantity that was of concern: As to the quantity of exercise, physicians argued that this should never exceed the limits of fatigue. Physician Timothy Bennett argued that once exercise had fatigued the body, any exercise taken after that point was considered 'immoderate', and risked excessive sweating and a 'wasting' of the body's strength.[94] Bennet explicitly strengthened his argument by referring to Santorio, whose experiments 'prove' that exercise continued after fatigue obstructed insensible perspiration.[95] In making these points, eighteenth-century physicians continued to rely on the authority of Santorio. 'Wasting' was a considerable concern to physicians of this period, and this concern continued to be prevalent in exercise advice throughout the eighteenth century and even into the nineteenth century. Ideas about what was the exact matter that was being wasted through exercise, or the manner in which this happened, shifted throughout the period under consideration, as medical discoveries were made, and theories changed. Yet the general fear of

[91] Santorio, *Medicina Statica* (version 1737), 263.

[92] Quincy, *Medicina Statica* (version 1712), 210.

[93] Byfield, *Two discourses*, 23.

[94] Timothy Bennet, *An essay on the gout; in which a method is propos'd to relieve the hereditary, and to cure the acquired … Such as have been reliev'd, or actually cur'd of the Gout by it* (London, printed for Richard Ford, 1734), 28–29, Eighteenth Century Collections Online.

[95] Ibid.

'running out of matter' was not new and had in fact existed since antiquity, although it does seem to have intensified by the continued publication of Santorio's work and the popularity of *Medicina statica*.

The mechanist's vision of the machine body, with its emphasis on minute fibres, eventually gave way to the model of the body as a reactive organism often referred to by historians as the 'sensible body', to reflect physicians' interest in the workings of the nervous system and how the body perceived through it. Despite this shift in focus, the theory that exercise could cause the body to run out of matter remained. Towards the end of the century, as the nerves were ascribed a central role in the motion and health of the body, physicians such as William Cullen (1710–1790) came to believe that it was in the nervous system that the generative source of life lay.[96] Although vitalism dominated the medical world by the late eighteenth century, much of the influence of iatromechanism endured, as has been pointed out by Hisao Ishizuka.[97] It is difficult to distinguish the different strands or gradations of the various medical frameworks, and many health guides contained advice based on a mixture of mechanist, vitalist, and humoural theories. This transition from the fibre body to the sensible body was anything but sudden, and many physicians found ways to combine the two doctrines, resulting in health advice that contained theories from these newer schools of thought as well as from much older ideas.[98] This explains why we still find many iatromechanical, Santorian, and even Galenic principles in vitalists' works, though many of these did wane in popularity.

Late eighteenth-century vitalists (physicians of a school of thought which argued that people possess a vital principle which can run out, at which point life ends) continued to warn readers of health guides to keep their exercise levels moderate, as the mechanists had done before them. John Trusler (1735–1820) argued: 'Now in the human system, there is a certain quantity only of living power that can be produced for the health of the body; so, that if we exhaust that power, we debilitate or weaken the body. A man, for example may walk, till he can walk no longer: it is,

[96] Cited in Sylvia de Renzi, "Women and Medicine", in *The Healing Arts: Health, Disease and Society in Europe, 1500–1800,* edited by Elmer, Peter. Manchester: Manchester University Press (2004), 189.

[97] Hisao Ishizuka, *Fiber, Medicine, and Culture in the British Enlightenment* (New York: Palgrave Macmillan, 2016). See especially xix, and chapter four.

[98] Ishizuka, *Fiber, Medicine, and Culture in the British Enlightenment*, 109–117.

therefore, necessary to know how far exercise is salutary'.[99] Vitalists' concept of this power, often referred to as 'Vital Power', is reminiscent of the theory of radical moisture. From antiquity, heat and moisture were already key concepts especially in relation to generation, ageing, dying, and health.[100] In the ninth and tenth century however, philosopher and physician Ibn Sina (unknown–1037), commonly known as Avicenna, ranked radical moisture as a constitutive element of life.[101] Avicenna argued that the body's innate heat could do nothing without radical moisture, which it needed as a fuel for a flame.[102] Following this theory, many later physicians believed everybody had a set quantity of moisture which gradually depleted over the course of one's lifespan, as the radical moisture, or vital power, was 'used up' in the process of repairing and maintaining the body.[103] Avicenna's concept of radical moisture echoed through eighteenth- and nineteenth-century health guides still, albeit with slightly different terms. 'The process of life', nineteenth-century physician Christopher Hufeland explained in 1829, could be considered 'a continued process of consumption: and its essence may be defined as an uninterrupted wasting and reparation of ourselves'. 'Life has often been compared to a flame', Hufeland continued, evoking Avicenna's famous metaphor, 'and indeed the operation in both is the same'.[104] That the image of the flame running out of oil and burning out had remained relevant centuries later indicates that it was a powerful one, but also that vitalists imagined the ageing process to function in a manner similar to traditional physicians.

Early modern health guides were often dedicated to finding ways of prolonging life; in the nineteenth century, many health guides still promised longevity on their frontispiece, such as Christopher Hufeland's *Art of Prolonging Life* (1829), T.J. Graham's *Methods of Improving Health and Prolonging Life* (1827), and James Adair's *Essay on Diet and Regimen as Indispensable to the Recovery and Enjoyment of Firm Health* (1804), to

[99] Trusler, *An easy way to prolong life*, 85.

[100] Gianna Pomata, "Innate Heat, radical Moisture and Generation", in *Reproduction: Antiquity to the Present Day*, eds. Nick Hopwood, Rebecca Fleming and Lauren Kassell (Cambridge: Cambridge University Press, 2018), 195–108.

[101] Ibid., 197–198.

[102] Ibid., 198.

[103] How much vital power a person had depends on what quantity of vital power was passed on by their parents at conception. Hufeland, 212; Stephen Pender, 'Heat and moisture, rhetoric and *spiritus*', *Intellectual History Review*, 24: 1, (2014), 90.

[104] Hufeland, *The Art of Prolonging Human Life*, 36.

mention but a few.[105] The quantity of vital power in a body was believed to be set and could not be added to, so the primary way to keep the flame lit for longer was by reducing the rate at which it consumed oil, or expended vital power. The only way to do this was by a 'due economization of the powers of vitality', as one nineteenth-century physician put it.[106] While vital power was used for all bodily functions, the best way to economise expenditure of vital power was to minimise those functions or motions which were voluntary. Vital power sustained loss by the body exerting its strength, Hufeland explained, and while sleep and nourishment could do much to repair the body without too much vital power being 'used up', violent or long continued exertions could exhaust the vital power completely. He explained: 'This is proved by common experience, as we find that after great exertion in walking, thinking &c. we become fatigued'.[107] Exercise of too violent a nature consumed the vital power and would result in insufficient power to repair the body.[108] These vitalists believed that the more intensely a being lives, the more its life will lose in extension.[109] Strengthening the organs was important to the uninterrupted flow of the fluids, but fatiguing exercise promoted no other effect than to make the organs drier, more rigid, and consequently sooner unfit for use. Therefore, instead of prolonging life, it would bring on premature old age and speedier dissolution.[110]

Roselyne Rey has argued that in this vitalist understanding of health and longevity, the individual was responsible for the regulation of their own sensibility, which they could over or under-stimulate, and consequently, for their own vital power.[111] Within this framework, she explained, '[d]isease is not only an internal disturbance, but an error in the distribution of sensibility, a mistake in the proportion', and argued that 'disease is then not, as in a mechanistic view, an interruption in the circulation of

[105] This focus on longevity can be seen throughout the eighteenth century, in all periods and all schools of medical thought.

[106] Medicus, *The Oracle of Health and Long Life, or Plain Rules for the Attainment and Preservation pf Sound Health and Vigorous Old Age; with Rational Instructions for Diet, Regimen &c.* (London: Printed for Sherwood, Gilbert & Piper, 1830), 9, www.archive.org.

[107] Hufeland, *The Art of Prolonging Human Life*, 34.

[108] Medicus, *The Oracle of Health and Long Life*, 123–124.

[109] Hufeland, *The Art of Prolonging Human Life*, 39.

[110] Ibid., 129.

[111] Roselyne Rey, "Vitalism, Disease and Society", in *Medicine in the Enlightenment*, ed. Roy Porter (Amsterdam, Editions Rodopi B.V., 1995), 279–280.

fluids', but instead, '[d]isease is a problem of management of resources, a problem of self-government'.[112] Her conclusion that vitalists were concerned with resource management is certainly true, as we have seen in the above examples. However, this concern does not necessarily constitute a break with the mechanics. There is a significant degree of continuation of mechanics' concerns with managing digestion and vitalists' fear of sweat disrupting the nourishment of the body.

Carol Houlihan Flynn has shown that medical theorists (and English novelists), whose works were committed to sustaining 'vitality' and prolonging life, warned against the dangers of 'solidification', which caused one's juices to grow stiff and solid.[113] This idea was already expressed in the early eighteenth-century work of Cheyne, who wrote:

> Time and Age will fix and stiffen our Solids. As in the greater world, the Quantity of the Fluids is Daily lessening and decreasing; so in our lesser world after a limited Time, the Appetite and Concoctions failing, the Fluids are lessened and spent on the continual repairs of the Solids, and thereby lose their Nature, and become firm and hard.[114]

The eighteenth-century concepts of vital power and sensibility of course differed from innate heat and radical moisture, but both theories subscribed to the idea that longevity and health were a matter of resource management. This, I have argued, also applies to early eighteenth-century mechanists, who focussed their concerns on the dangers of running out of sufficient matter to nourish the solids. Rather than seeing this aspect of vitalism as a departure from older views of the body, I suggest that it is in fact a continuation of the same idea. Whether the emphasis was on solids, nourishing matter, moisture, heat, or power, what eighteenth-century physicians (and, in fact, also physicians preceding them) shared was the idea that one could run out of 'matter' and that exercise, if not carefully moderated, could hasten the diminution of this matter considerably.

Despite the increased focus on vital power, discussions about perspiration and sweat continued in late eighteenth-century medical texts and health guides, albeit with an emphasis on the senses, typical for vitalism,

[112] Ibid., 280–281.

[113] Carol Houlihan Flynn, "Running out of Matter: The Body Exercised in Eighteenth-Century Fiction", in *The Languages of Psyche*, ed. G. S. Rousseau (Oakland: University of California Press 1990), 150.

[114] Cheyne, *An Essay of health and long life*, 220–221.

rather than minute vessels or the mechanical processes of the body. We can find ideas about perspiration expressed in their new context in the writings of Francis de Valagin (1724–1805), who considered the effects of emotions on circulation and perspiration:

> Joy is, when moderate, the most salutary of all the passions…it increases the circulation of the blood and of the animal spirits; gives Vigour to all the Faculties; nourishes and moistens the whole habit of the body.[115] Sorrow is the direct opposite of Joy, a passion hardly serviceable to any body, except to those who are very fat, drowsy, and sleepy, or those who are overjoyed … As Joy increases perspiration, so sorrow obstructs it; and from an obstructed perspiration arises grief and fearfulness. Sorrow turns the face pale, renders the nervous force languid, and diminishes all vital motion.[116]

The mind and body were believed to be intimately connected and physicians increasingly emphasised that exercise should be pleasing to the mind, as well as to the body.[117] Exercises such as walking or riding were recommended for their effect on both the body and the mind: 'the first obtains a greater tone of vigour, and is sustained in a more uniform temperature; the second, a greater serenity than can be obtained by any other method of exercise'.[118] Exercises which offered pleasing scenery, or which could be done in the company of friends, were believed to be especially beneficial to health.[119] While the 'universal motion' that exercises such as horseback riding brought continued to be lauded, the consideration of the effects on the fibres was not the only or even the primary factor physicians considered when dispensing exercise advice. From the second half of the eighteenth century, the state of the mind and its effects on insensible perspiration featured high on the list of considerations. New vitalism did little to affect ideas about the healthful limits of exercise, however. Exercises were still to be moderate in their intensity and duration in order to prevent sweating and exhausting vital power.

[115] Francis De Valagin, *A treatise on diet, or, the management of human life; by physicians called the six non-naturals* (London: Printed for the author by J. and W. Oliver, 1768), 321, Eighteenth Century Collections Online.

[116] Ibid., 325–326.

[117] Ibid., 306.

[118] Medicus, *The Oracle of Health and Long Life*, 115.

[119] Ibid., 127; De Valagin, *A treatise on diet, or, the management of human life*, 215.

Although gradual, there are some indications that fears concerning sweat and wasting started to abate in the nineteenth century. Shigehisa Kuriyama argued that the management of insensible perspiration remained a topic of great concern until around the 1840s, when physicians conceived of the principle of the conservation of energy.[120] As she points out, the idea that the *ingesta* could be converted into fuel for the body had disproven Santorio's theories as much of the difference between the *ingesta* and *excreta* was now attributed to energy, rather than to insensible perspiration. This reduced the role of perspiration in the digestive process significantly from what Santorio had argued. Recently, James Kennaway and Rina Knoeff argued that already in the eighteenth century ideas of moderation in exercise changed to the extent that 'getting sweaty and out of breath was not a problem as long as one took proper breaks between exercises', due to the eighteenth-century emphasis on the importance of the strength of the fibres of the body.[121] As we have seen however, physicians continued to express concerns about sweat until the very end of the eighteenth century, including those physicians who also acknowledged the importance of keeping the fibres of the body strong. Indeed, the model of the fibre body co-existed with the theories that argued that sweat was a preternatural and violent discharge which deprived the fibres of the body of nourishment, thus weakening them. It certainly seems that the fear of sweating ceased to be quite so prominent by the early nineteenth century, however. Graham, for example, argued in 1827 that most people 'are much more prone to err on the side of deficiency, than on that of excess' when it came to establishing the right quantity of exercise.[122]

Conclusion

The benefits and dangers of exercise took up considerable space in the health literature of the long eighteenth century. Recommended on the one hand for its ability to stimulate all the bodily processes from digestion to circulation, exercise was also feared for its perceived ability to waste the body. As I have demonstrated, this fear was especially centred around

[120] Kuriyama, "The Forgotten Fear of Excrement", 431.

[121] Kennaway and Knoeff, "For it is the debilitating fibres that exercise restores", 113.

[122] T. J. Graham, *Sure Methods of Improving Health and Prolonging Life; Or, A Treatise on the Art of Living Long and Comfortably by Regulating The Diet and Regimen* (London: Published for the author by Simpkin and Marshall, 1827), 241.

sweat, which many perceived to be a sure sign that the exercises performed were excessive. The idea that sweating was detrimental to health and the fact that sweaty exercise was therefore strongly discouraged is very much at odds with what our modern society views as healthy exercise, and for that reason alone deserves the attention of historians of health and the body. In this chapter, I have demonstrated that many authors of health guides and other medical texts strictly differentiated between sweat and insensible perspiration, lauding the health benefits of the latter while warning against the detrimental effects to health of the former. Although these ideas originated in ancient medicine, I have suggested that the translation and re-publication of Santorio's aphorisms gave a new impetus to the notion that sweating was detrimental to health. Thus Santorio's argument that sweating disrupted the flow of insensible perspiration, which was not only the more considerable, but also the most perfect excretion, was part of the early eighteenth-century health guides and shaped the exercise advice found in them. Eighteenth-century health guides recommended that people exercise 'ad ruborem, non sudorem', which resulted in an exercise culture which revolved around carefully managing sweat and insensible perspiration.

All health advice did not apply to all bodies equally, as different people had different exercise needs. In the seventeenth century, physicians argued that certain constitutions had different exercise needs from others and that some bodies, especially those of a 'phlegmatick' constitution, needed exercise even more than others. This idea can still be found in some eighteenth-century treatises, although increasingly physicians recommended exercise based around specific illnesses rather than constitutions. Those who were morbidly corpulent were sometimes advised that they should sweat, despite the knowledge that sweating could cause weakness of the solid fibres and disruption to the flow of insensible perspiration. Physicians also differentiated exercise advice according to gender, although it seems this was based on assumptions about women's lifestyles rather than their physiology. Women, similarly to fat people, were understood to often suffer from plethora, yet in the former the plethora was not an illness, but rather a natural part of their physiology. This set exercise advice aimed at women apart from exercise advice aimed at corpulent bodies. Physicians explicitly stated that it was only when 'tis to give Relief from some greater Evil', such as a fever, or morbid corpulency, that sweaty exercise was permitted. Age also determined the nature of the exercises advised. I have suggested that this is partly due to the idea that old people could not exercise as vigorously as younger people, but also because old people were drier and therefore more likely to suffer after sweating.

As medical theories shifted over the course of the eighteenth century with the onset of solidism and new vitalism, some of Santorio's theories were challenged. However, Santorio's doctrine of insensible perspiration remained popular. I have argued that the more general idea that the body could 'run out of matter' survived these shifts, and that exercise advice continued to contain warnings against exercising excessively so as to prevent wasting. The new emphasis on the solids increased the interest in exercise advice, but also sharpened fears of overdoing exercise and thus damaging the solids that people were advised to strengthen. Authors of health guides therefore recommended exercises which provided constant, low-impact movements to the whole body and which were not so intense as to cause sweating. In the late eighteenth century, the language used by physicians changed, but the warning that exercise ought not to produce sweat stayed the same.

This is not to say that the idea of exercising 'ad ruborem, non sudorem' was the defining aspect of eighteenth-century exercise advice, nor even necessarily the most important factor to readers of these guides. As many historians have already pointed out, the perceived threat to health posed by suppressed perspiration was pervasive throughout health literature of this period, and guidance on how to prevent suppressions was even more prevalent than advice on how to moderate excretions. Nevertheless, fear of suppressed evacuations does not provide us with the full picture of eighteenth-century exercise concerns, as the health guides of this period almost always included warnings about excessive or violent exercise, which should not be ignored. This chapter has demonstrated that the old saying 'ad ruborem non sudorem' was an important part of eighteenth-century exercise advice and practices; understanding where this idea came from, how it developed as medical theories changed, and why it eventually disappeared from health advice will help us develop a fuller understanding of eighteenth-century theories about bodies and health.

CHAPTER 6

A Weighty Matter: Understanding Corpulency and Body Weighing Practices in Eighteenth-Century Britain

Body weighing as a practice of health maintenance was first introduced by Paduan physician Santorio Santori in the late sixteenth century as part of an experiment to measure how much the body excreted through insensible perspiration in a 24-hour period.[1] The practice of weighing oneself became increasingly common in Britain over the course of the eighteenth century, though extant evidence suggests that the practice remained largely restricted to the urban elite. In this chapter, I will argue that to understand body weighing as it was practised in the seventeenth and eighteenth centuries, we must first disconnect weight from fatness, no matter how intrinsically connected the concepts seem today. There is a tendency to apply modern terms such as 'weight-loss' and 'overweight' to what eighteenth-century people called 'reducing' and 'corpulency'.[2] Our

[1] As discussed in Chap. 1, the results of this weighing experiment were published by Santorio in his most famous work, the *Ars de statica medicina* (Venice, 1614). For a comprehensive overview of Santorio, his work, and his influence on physicians and chemists including Robert Boyle (1627–1691), Carl Linnaeus (1707–1778), and Antione Lavoisier (1743–1794), see Jonathan Barry and Fabrizio Bigotti, *Santorio Santori and the Emergence of Quantified Medicine, 1614–1790: Corpulscularianism, Technology and Experimentation* (London: Palgrave Macmillan Cham, 2022).

[2] Christopher Forth does this occasionally in his book, in which he explains that his use of the word 'overweight' is a means for him 'to convey the impressions of various periods'. Christopher E. Forth, *Fat: A Cultural History of the Stuff of Life* (London: Reaktion Books,

M. Wübbels, *Quantification and Measurement in British Health Advice, c. 1670–1830*, Palgrave Studies in Medieval and Early Modern Medicine,
https://doi.org/10.1007/978-3-032-32760-4_6

tendency to assume diets to be attempts at weight loss is understandable but obscures contemporary understandings of bodies and health. In the eighteenth century, corpulency was still defined and diagnosed with reference to a well-functioning body and the capacity for exercise and day-to-day activities, rather than weight. Body weighing did not feature in medical discussions about corpulency, and corpulence was not mentioned in texts discussing the benefits of body weighing. It is not until the end of the eighteenth century that we first see a connection being made between high body weight and fatness. Popular interest in body weighing may have been owing to translations and new editions of Santorio's work, which were published over the course of the early eighteenth century, as well as to eighteenth-century re-trials of Santorio's weighing experiment.[3] However, as this chapter will demonstrate, the practice of weighing oneself in the eighteenth century had deviated from Santorio's methods in the early seventeenth century, and so we must be careful not to assume that practice of body weighing is a linear development from Santorio's first experiment.

Focussing on Santorio's first experiment and the eighteenth-century re-trials, this chapter will explain the motivations behind the weighing experiments undertaken by physicians. These weighing experiments have already received considerable attention from historians, especially Lucia Dacome. In her article 'Living with the chair', Dacome discusses the development of the practice in the eighteenth century, focussing especially on how this form of self-measuring was undertaken by physicians. Dacome's careful analysis lays bare the difficulties physicians sometimes had in convincing others of the medical authority of the weighing machine

2019), 285, and James Kennaway and Rina Knoeff argue that in the eighteenth century, exercise was recommended especially to those who were 'overweight'—but the concept of overweight did not yet exists at that time. James Kennaway and Rina Knoeff, '"For it is the debilitating fibres that exercise restores": Movement, morality and moderation in eighteenth-century medical advice literature', in James Kennaway and Rina Knoeff (eds.), *Lifestyle and Medicine in the Enlightenment: the six non-naturals in the long eighteenth century* (Abingdon: Routledge 2020), 112.

[3] The first English editions to appear were by Cole (1663) and Davies (1676). The *Ars de statica medicina* was later also translated by John Quincy, as *Medicina Statica, being the Aphorisms of Sanctorius, translated into English, with large Explanations* (London, 1712). Quincy's translation reached multiple editions over the course of the early eighteenth century, hence that I refer to *the Ars de statica medicina* as *Medicina statica*.

and self-experiment.[4] In 'Balancing Acts', Dacome continues her investigation of eighteenth-century weighing practices by examining the depiction of a man sitting in a weighing chair, which often accompanied the various editions and translation of Santorio's *Medicina statica*.[5] Dacome explains how the image (see cover image of this book) imparted a sense of continuity to the practice amid changing medical theories and increasing criticism. As Dacome's work has shown us, the practice of body weighing in medical circles was highly contested and fraught with uncertainties as to how a weighing experiment ought to be undertaken and what authority the results of such an experiment held.

This chapter will further our investigation of eighteenth-century body weighing practices by asking not only why body weighing became increasingly practised over the course of the eighteenth century, but also how and why these eighteenth-century body weighing practices differed from Santorio's experiment. I assess late seventeenth- and eighteenth-century literature on corpulency, as well as the earlier literature written by Santorio that influenced these later authors and those physicians who partook in weight-watching experiments. I will first highlight the complex relationship between eighteenth-century discussions about fatness and the practice of body weighing. As I will demonstrate, these two topics were discussed in wholly different contexts, until the late eighteenth century, when we start to find evidence of connections being made between fatness and high body weight. Such a discussion is important to the history of body weighing, as our modern tendency to connect body weight with fatness skews our ability to understand eighteenth-century motivations for keeping track of body weight. It also helps us understand the significance of the changes that we see towards the end of the eighteenth century, when we see the first evidence of the concept of 'overweight' in a letter exchanged by physician William Cullen (1710–1790) and one of his patients.

I will then shift the focus to Santorio's weighing experiment and how physicians presented the practice of body weighing in eighteenth-century medical literature. I will discuss especially physicians' attempts to dissuade readers from embarking on Santorian-esque weighing experiments. As this

[4] Lucia Dacome, 'Living With the Chair: Private Excreta, Collective Health and Medical Authority in the Eighteenth Century', *History of Science*, 2001, 39, 467–500.

[5] Lucia Dacome, 'Balancing Acts: Picturing perspiration in the long eighteenth century', *Studies in History and Philosophy of Biological and Biomedical Sciences*, 2012, 43, 379–391.

chapter will explain, the purpose of a weighing experiment was to learn what caused shifts in the body's weight in a 24-hour period, so that the experimenter might learn how to avoid significant fluctuations in weight, which was considered a sign of ill health. Tracing minute changes in body weight required frequent weighing throughout the day and careful consideration for all that could affect body weight, such as food, exercise, and even the humidity and temperature of the environment the weighing experiment was conducted in. For that reason, eighteenth-century physicians did not recommend that non-medical people adopt the practice. Body weighing was deliberately restricted to the medical circles, where it continued to be practised as an intricate experiment undertaken in a controlled environment until the late eighteenth century. We will see in the last section that physicians were ultimately unsuccessful in dissuading non-medical people from weighing themselves. However, I argue that physicians' warnings about weighing experiments nevertheless may have influenced how non-medical people approached the practice of body weighing in the second half of the eighteenth century.

In the last section of this chapter, I demonstrate how late eighteenth-century weighing practices of non-medical people deviated from the practices of Santorio and early eighteenth-century experimenters. Here, I assess the data from a still largely untapped resource, namely, the weighing ledgers of the eighteenth-century grocery shop then located at The Sign of The Coffee Mill (henceforth: Coffee Mill), in St. James Street in London, where the upper and middling sorts of London society and those visiting London could go to be weighed from the mid-eighteenth century onwards.[6] Containing the entries of thousands of weightwatchers over the course of the late eighteenth and nineteenth centuries, these ledgers provide a unique insight into weighing practices outside of the medical circles. The patterns showing in the data from the ledgers suggest that these weightwatchers engaged in the practice in a manner which is reminiscent of, but considerably different than, Santorio's weighing experiment. This chapter argues that the late eighteenth-century instances of body weighing should not be considered a linear development from seventeenth-century body weighing practices, but as a practice uniquely informed by medical theories and practical limitations, and even local customs.

[6] The first, to my knowledge, to mention these ledgers in the context of medical history is Dacome, in 'Living with the Chair', 488–489.

Corpulency: A Matter of Great Weight?

As historians such as Lucia Dacome, Michael Stolberg, Georges Vigarello, Ken Albala, Pat Rogers, Sarah Toulalan, and Cristopher E. Forth have demonstrated, corpulency was a topic that was increasingly written about over the course of the seventeenth and eighteenth centuries.[7] The first English book dedicated to this subject was Thomas Short's 1727 publication *A Discourse Concerning the Causes and Effects of Corpulency.* Malcolm Flemyng published his *Discourse* in 1760, and the other two English texts, William Wadd's *Cursory Remarks on Corpulence* and his *Comments on Corpulency,* date from the beginning of the nineteenth century.[8] However, it was also discussed in many other health-related publications, especially in texts advising readers on how to live long and healthily.

Having identified the eighteenth century as a pivotal point in the history of fat as a health concern, some historians have understandably considered the eighteenth century as the beginning of the history of modern weight-watching practices.[9] In 'Weight loss in the Age of Reason', Ken

[7] Lucia Dacome, 'Useless and Pernicious Matter: Corpulence in eighteenth-century England', in Christopher E. Forth and Ana Carden-Coyne, eds, *Cultures of the Abdomen: diet, digestion, and fat in the modern world*, eds (Basingstoke: Palgrave Macmillan, 2005); Michael Stolberg, "Abhorreas pinguedinem': Fat and obesity in early modern medicine (c.1500–1750)', *Studies in History and Philosophy of Biological and Biomedical Sciences*, 2011, doi:10.1016/j.shpsc.2011.10.029; Georges Vigarello, *The Metamorphoses of Fat: A History of Obesity* (Columbia: Columbia University Press, 2013); Pat Rogers, 'Fat is a Fictional Issue: The Novel and the Rise of Weight-Watching', in Elena Levy-Navarro, ed, *Historicizing Fat in Anglo-American Culture* (Columbus: Ohio State University Press, 2010); Ken Albala, 'Weight Loss in the Age of Reason', in Christopher E. Forth and Aana Carden-Coyne, eds, *Cultures of the Abdomen: diet, digestion, and fat in the modern world* (Basingstoke: Palgrave Macmillan, 2005); Sarah Toulalan, "To[o] much eating stifles the child': fat bodies and reproduction in early modern England', *Hist Res*, 2014, 87, 65–93; Forth, *Fat: A Cultural History of the Stuff of Life*.

[8] Thomas Short, *A Discourse Concerning the Causes and Effects of Corpulency: Together with the Method for Its Prevention and Cure* (London: J. Roberts, 1727); Malcolm Flemyng, *A Discourse on the Nature, Causes, and Cure of Corpulency* (London: L. Davis and C. Reymers, 1760); William Wadd, *Cursory Remarks on Corpulence; or, Obesity considered a Disease: with a Critical Examination of Ancient and Modern Opinions, relative to its Causes and Cure* (London: J. Callow, 1810); William Wadd, *Comments on Corpulency, Lineaments on Leanness, Mems on Diet and Dietetics* (London: John Ebers & Co., 1829).

[9] As is evident from Roy Porter arguing that 'weight loss' became something of an obsession in the eighteenth century, and Pat Rogers naming his chapter on fat in eighteenth-century literature 'The Novel and the Rise of Weight-Watching'. Roy Porter, *Flesh in the Age of Reason* (London: Penguin Books, 2003), 234; Rogers, 'Fat is a Fictional Issue'.

Albala argued that 'by calling attention to the dangers of excess weight', medical professionals of the seventeenth and eighteenth centuries had been responsible for a 'gradual shift in meaning of the word diet from a general programme of health to a regimen specifically designed to cure excess weight'.[10] However, eighteenth-century medical professionals were not calling attention to the danger of excess weight, but rather to the dangers of excess blood, or excess fat/oil in the body, depending on the particular period or school of thought to which the medical professional belonged. Discussions about the dangers of corpulency (a term used in the eighteenth century to describe a body type or body state characterised as being fatter than other body types or states) did not usually refer to body weight. For the majority of the eighteenth century, body weight did not yet play a role in the medicalisation of corpulency.

Today, body weight plays a tremendous role in our defining, diagnosing, and curing of excess fat. The World Health Organization (WHO) applies the term 'overweight' to 'abnormal or excessive fat accumulation', denoting our modern tendency to measure and define 'abnormal or excessive' levels of fatness through body weighing.[11] Corpulency, however, was defined very differently. Physician Thomas Short (1690–1772) described it as 'that gross Habit of Body', which had increased in bulk due to 'a Great Plenty of Blood stor'd with oily Parts', to the extent that it would 'hinder, prejudice, or render the Performance of the Actions of Life uneasy or painful'.[12] It was this disruption to the bodily functions which made corpulency a disease, and the primary factor by which medical professionals of the eighteenth century measured and classified the corpulent body. The reason for this can be found in early modern understandings of the concept of health. Dutch chemist and physician Herman Boerhaave (1668–1738) described the contemporary idea of health as: 'whoever can exercise the proper actions of life, with ease, pleasure, and some certainty of time, may well be esteemed to have a sound and healthy body'.[13] The opposite of health was the inability to perform these actions of life, and it is on this understanding of health and disease that the medicalisation of corpulent bodies was based. Scottish physician Malcolm Flemyng

[10] Albala, 'Weight Loss in the Age of Reason', 169.

[11] Taken from the World Health Organization page on Obesity and Overweight: https://www.who.int/news-room/fact-sheets/detail/obesity-and-overweight

[12] Short, *A Discourse Concerning the Causes and Effects of Corpulency*, 9–10

[13] Ibid., p. 7.

(unknown–1764) stated that 'corpulency may be reckoned a disease, as it in some measure obstructs the free exercise of the functions', and William Cullen concluded that 'there is a certain degree of it, which will be generally allowed to be a disease; as when it renders persons, from a difficult respiration, uneasy in themselves, and from the inability of exercise, unfit for discharging the duties of life to others'.[14] Only on some occasions did physicians mention the measured or estimated weight of extremely fat individuals, but even in these cases, body weight was not used for determining whether these individuals were corpulent.

Instead, patients were diagnosed based on their physical ability to perform day-to-day tasks and physical exercises. When medical professionals assessed their corpulent patients, they looked for signs of reduced physical abilities, as can be seen in this case from English physician John Fothergill's (1712–1780) *Medical Observations* (1776):

> A country tradesman, aged about thirty ... and very fat, applied to me for assistance. He complained of perpetual drowsiness and inactivity. His countenance was almost livid; and such a degree of somnolency attended him, that he could scarce keep awake whilst he described his situation. In other respects he was well.[15]

In another case Fothergill described a young woman, 'about twenty-three years of age, of a low stature, and very fat', who applied to him for assistance because she had 'great difficulty of breathing', as well as an 'incapacity for any exercise'.[16] But while reduced physical ability was associated with corpulency, not every fat person was presumed physically impaired. In an account of the then famous fat man Mr. Bright, we find that its author did not ascribe to Bright any of the characteristics that were ascribed to Fothergill's patients:

> He was always fat from a child, and yet very strong and active, and used a great deal of exercise, both when a boy, and after he became a man, which he continued to do till within the last two or three years of his life, when he

[14] Flemyng, *A Discourse on the Nature, Causes, and Cure of Corpulency*, 1; William Cullen, *First Lines of the practice of physic*, Vol. 4 (Edinburgh: Printed for C. Elliott, 1784), 218–219.

[15] John Fothergill, 'Case of an Angina Pectoris', in John Elliot, ed, *A Complete Collection of the Medical and Philosophical works of John Fothergill* (London: Printed for J. Walker, 1781), 525.

[16] Ibid., 526.

> became too unwieldy. He could walk about very well, and nimbly too, having great strength of muscles; and could not only ride on horseback, but would sometimes gallop after he was grown to between 30 or 40 stones weight.[17]

This description renders him essentially free from the label of that state of morbid corpulency Short described. Indeed, while he weighed anywhere between 30 and 40 stone, Mr. Bright was described as perfectly healthy, because he continued to be physically active and able to get about without difficulty. The same applies to a similar— although much later— case, concerning Daniel Lambert from Leicester. His story was printed a pamphlet in 1818, with much emphasis on his physical state:

> Lambert scarcely knew what it was to be ailing or indisposed. His temperance, no doubt, contributed towards his uninterrupted flow of health. He had a fine, powerful and melodious voice. It was a strong tenor, unlike that of a fat man, light, and unembarrassed, and the articulation perfectly clear. He never felt any pain in his progress towards his extraordinary bulk, but increased gradually and imperceptibly.[18]

Lambert's weight—more than 52 stone at the time of his death—was extreme, yet the author of this text made a point of arguing that Lambert was the very picture of health, distinguishing him from the medical cases of corpulency that we read about in the works of Fothergill.

Corpulency was therefore not simply an older term for what we today call obesity or overweight. While we classify it as a medical condition, the term corpulency was used to indicate an innocuous body type, as well as a morbid state. An assessment of 41 cases from the casebook of physician John Bayly (1762–1764) shows that Bayly knew that several of his patients were bulkier than most men or women.[19] Yet he did not include in his notes any indication that he advised them to reduce themselves, which

[17] T. Coe, 'A Letter from Dr. T Coe, Physician at Chelmsford in Essex, to Dr. Cromwell Mortimer, Secr. R. S. Concerning Mr Bright, the Fat Man at Malden in Essex', *Phi. Trans. R. Soc.*, 1752, 46, 189.

[18] Anon., *The Life of that wonderful and extraordinarily heavy man, Daniel Lambert, from his birth to the moment of his dissolution; with an account of men noted for their corpulency, and other interesting matter* (New York: Samuel Wood & Sons, 1818), 26–27. https://wellcomecollection.org/works/xycvjyp8.

[19] Case Book of John Bayly, 27 Sept. 1762–1 Sept. 1764, (MS105). Royal College of Physicians Archives.

suggests that he did not regard corpulency as a health issue *in itself* that required treating. Neither did he note any of his patients' body weights, despite describing several of them as corpulent. While this may have been reflective of the limited availability of scales, it is apparent that body weight was unimportant in reaching a diagnosis of corpulency.

Patients who were physically suffering under their excess fat were advised to 'reduce' themselves. Physicians advised not a reduction of weight but a reduction of the body's 'plethora' of excess blood or oil or fat residing in the body. It was believed that the body could naturally excrete excess oil or fat through the excretory system (stool, urine, and sweat). It naturally followed that if these excretions could be increased, the quantity of oil or fat in the body could be diminished.[20] Flemyng was certain that frequent purging, through diuretics, or induced diarrhoea or vomiting, 'would no doubt be a speedy and effectual means of reducing corpulency'.[21] Bath physician George Cheyne (1672–1743) explained how he himself managed to reduce his own corpulency by increasing his evacuations (especially through vomiting and sweating) and decreasing his dietary intake, until his 'blood had entirely lost its size and all the former habit was wasted, wore away, and discharged'.[22] Generally, the success of such a reduction was measured by how one felt afterwards. A diminution of the plethora which had oppressed the body's functions was expected to result in recovery of those functions and a general sense of well-being. There are ample instances of this idea in the literature concerning corpulency and dieting. Cheyne, for example, described himself as feeling 'lank, fleet and nimble' after his reduction.[23] Weight was used occasionally to measure a reduction of bulk: Cheyne added that he had lost 'almost one third in weight and measure', though it does not appear that losing weight was Cheyne's objective.[24] The fact that Cheyne does not give his starting and finishing weight suggests that Cheyne valued weight primarily as a means to quantify his reduction of fat, but that he attributed no value to either knowing or sharing his current or previous total body weight.

Occasionally, body weight was mentioned in case studies of extremely heavy individuals, which featured in medical treatises on corpulency and in

[20] Flemyng, *A Discourse on the Nature, Causes, and Cure of Corpulency*, 15.

[21] Ibid., 16.

[22] George Cheyne, *The English malady: or, a treatise of nervous diseases of all kinds* (London: Printed for G. Strahan, 1733), 353–354, Eighteenth Century Collections Online.

[23] Ibid., 338.

[24] Ibid., 354.

the scientific journals. Thomas Short, in his *Discourse Concerning the Causes and Effects of Corpulency* (1727), provided his readers with some historical examples of extremely corpulent people, reporting that 'Sennertus saw a man who weighed 400 lib. weight, and a woman who weighed 420 lib. Nay, I myself saw a young lady, who died of corpulency in the 25th year of her age, who weighed above 500 lib.'[25] Short clearly expected his readership to be able to imagine how these weights translated into body sizes and shapes. While this suggests that people recognised the relationship between fat bodies and high body weight, it is important to acknowledge that these figures are very extreme and that Short probably included them to emphasise the prodigiousness of the young lady he described as 'a Monster in Nature for Bulk'.[26] The cases of Bright and Lambert (published in *Philosophical Transactions of the Royal Society* and in *Medical and Physical Journal*, respectively) also contained details about their body weight. Coe provided his readership with weight recordings from Bright's childhood to his death, allowing the reader to imagine the course of Bright's ballooning:

> what his exact weight was at the time of his death, cannot be told; but … if we take the same proportion, by which he has increased for many years on average, viz. of about 2 stones a year, and only allow 4 pounds addition for last year, on account of his moving about but very little, while he continued to eat and drink as before this will bring him to 44 stones or 616 pounds neat weight.[27]

The *Medical and Physical Journal* reported of Daniel Lambert that his weight was 'upwards of fifty stone, fourteen pounds to the stone'.[28] Such case histories of the extremely heavy should perhaps be considered in the context of a wider culture of reporting on extreme bodies rather than as part of discussions about corpulency, as they did not contain warnings about the dangers of corpulency but sensationalised the unusual weight, size, and lives of these men.[29] The authors of both Bright's and Lambert's

[25] Short, *A Discourse Concerning the Causes and Effects of Corpulency*, 9.

[26] Ibid.

[27] Coe, "Concerning Mr Bright, the Fat Man at Malden in Essex", 190.

[28] 'Medical and Physical Intelligence', *Medical and Physical Journal*, June 1806, 15 (88), 582–584, 583.

[29] See, for example, Jan Bondeson. *Freaks: The Pig-Faced Lady of Manchester Square and Other Medical Marvels*. (Stroud: Tempus Publishing, 2006), in which the case of Daniel Lambert is discussed in chapter four.

accounts seemed to purposefully deny any similarities between these 'wonderful and extraordinarily heavy' men and the corpulent patients discussed by Short and Flemyng, by emphasising their physical abilities, such as Bright's skill on horseback and Lambert's powerful voice, which was 'unlike that of a fat man'.[30] These examples show that extreme weight was not yet intrinsically connected to corpulency, which is of course a stark contrast to our modern use of the term obesity, the diagnosis of which is in part determined by body weight.

There is however evidence to suggest that in medical circles at least, ideas around body weight were changing towards the end of the eighteenth century. Just 30 years after Coe praised the strength and health of 40-stone weighing Bright, a physician's daughter named Jane Webster wrote to Edinburgh physician William Cullen to ask him for advice regarding weight loss.[31] Webster's letter presents us with a unique case, and not just for the fact that she, as a woman especially, had weighed herself with some regularity. What is perhaps most remarkable about this letter is that Webster expressed that she had an idea of what would be a 'moderate' weight for her, namely, 12 stone instead of her '15& odd pounds'.[32] This idea of there being a 'moderate' size was also shared by Cullen himself, who confirmed in his response to Webster, 'For a woman of middle size, any thing above 12 stone is too much'.[33] It appears then that Webster and Cullen both had a fairly precise idea of what constituted a healthy weight range for women, and presumably also for men, although there appears to be no similar conversation with a male patient in Cullen's correspondence. The idea of there being a measurable range wherein one must remain to be healthy was nothing new, as we will discuss in the next section, but Cullen and Webster's exchange is the first known instance where the boundaries of health are determined by body weight. Cullen's statement that any weight above 12 stone is too much is indicative of him, and presumably other medical professionals in this period of time, having developed the concept of *over*weight—although that term was not yet used. While this source is the only one I have found in which weight is discussed

[30] Anon., *The Life of that wonderful and extraordinarily heavy man, Daniel Lambert*, 26–27.

[31] Jane Webster to William Cullen, 11 September 1780 (Doc. ID: 1914), The Consultation Letters of Dr William Cullen (1710–1790) at the Royal College of Physicians at Edinburg, accessible online via The Cullen Project, http://www.cullenproject.ac.uk.

[32] Ibid.

[33] William Cullen to Jane Webster, 16 September 1780 (Doc. ID: 1021), The Consultation letters of William Cullen.

in this manner, it is an indication that by the end of the eighteenth century, physicians had started to espouse the idea that a certain weight could be 'too much' and a medical problem in itself, regardless of whether the patient felt physically affected by this weight.

Experimenting with Weight

Discussions about the theory and practice of weighing oneself did find expression in the medical literature of the early eighteenth century, albeit in a separate context than that of medical debates of corpulency. The use of scales to weigh the human body in a medical context was first introduced by Santorio, who weighed himself and others daily over a long period of time as part of an experiment and published the result of this experiment in his most famous work, the *Medicina statica*, which consisted of a series of aphorisms arranged according to the six non-naturals, with the addition of a chapter on insensible perspiration. It was in the chapter about insensible perspiration ('perspiratio insensibilis') especially that Santorio discussed the role of body weight in relation to health maintenance.

Insensible perspiration was described by Santorio as that perspiration which was too minor to be perceived by us and which passed through either the pores of the body or the mouth as steam or vapour.[34] According to Santorio's measurements, the amount of insensible perspiration which a person excreted amounted to about half a pound in a 24-hour period, which made insensible perspiration the body's most considerable excretion.[35] Santorio argued that it was therefore imperative to know and understand this particular excretion in order to maintain a perfect balance of our *ingesta* (the sum of ingested food and drink) and *excreta* (the sum of all excretions, including stool, urine, sweat, and insensible perspiration).[36] The mechanism with which Santorio had weighed himself was of his own invention and was precise enough in its readings to perceive minute changes in body weight, which had hitherto not been possible. With such a device, Santorio could keep track of exactly how much his body weight shifted throughout the day, and measure the effects of eating and drinking, urinating, bowel movement, exercising, and perspiring on his body

[34] Santorio, *Medicina statica* (version 1737), 45.
[35] Ibid., 44.
[36] Ibid.

weight. Santorio's invention made it possible for Santorio to perceive exactly how much his body had excreted and, therefore, also how much he needed to consume to make up for that loss. If he had mismanaged the balance between his *ingesta* and *excreta*, his weighing device would make him aware of a change in his body weight. In doing so, Santorio explicitly forged a link between body weight and health.

Santorio's body weighing experiments were later re-tried, republished, and re-discussed over the course of the early eighteenth century by physicians such as John Floyer (1649–1738), George Cheyne (1671–1743), James Keill (1673–1719), Bryan Robinson (1680–1754), and John Quincy (1683?–1723).[37] The medical school of thought known as iatromechanism, especially influential in the late seventeenth and early eighteenth centuries, favoured static approaches to health management, and it is in the context of this culture of static medicine that the weighing experiments of Santorio were revisited, almost a hundred years after they were first published in Santorio's *Medicina statica*. Just as Santorio, eighteenth-century physicians weighed not to know the absolute weight of the body, but rather to monitor and explain the daily fluctuations in weight caused by ingesting and excreting.

A consistent and careful weighing experiment could help the physician establish exactly how much was lost daily through excretions and how much the body would need to ingest to make up for that loss. The purpose of the weighing experiment was to learn how to maintain a consistent body weight by returning the body to the exact same weight it had been at the start of a 24-hour period, as physician Edward Barry explains here:

[37] Dacome points out that Santorio's medical doctrine was re-appropriated especially by iatromechanical physicians who pursued 'Newtonian' medicine, which included George Cheyne, James Keill, and John Quincy, in 'Living with the Chair', 471–472. For further information about Newtonian medicine and its popularity among late seventeenth- and early eighteenth-century physicians, see Anita Guerrini, 'Archibald Pitcairne and Newtonian Medicine', *Medical History*, 1987, 31 (1), 70–83, and Anita Guerrini, *Obesity and Depression in the Enlightenment: The Life and Times of George Cheyne* (Oklahoma: University of Oklahoma Press, 2000), esp. chapter 2. See also James Keill, *Essays on the several parts of the animal oeconomy* (London: Printed for George Strahan, 1717); Bryan Robinson, *A dissertation on the food and discharges of human bodies* (Dublin: Printed for S. Powell, 1747), Eighteenth Century Collections Online; William Stark, *The works of the late William Stark, M.D. consisting of clinical and anatomical observations, with experiments, dietetical and statical, revised and published from his original Mss by James Carmichael Smyth* (London: Printed for J. Johnson, 1788), Eighteenth Century Collections Online.

> There is a certain degree of weight in all bodies, under which they enjoy the most perfect health that such are capable of attaining to … if such a Body therefore returns once in twenty-four Hours to this Standard Weight of Health, it will continue free from any habitual Disorder.[38]

From a modern perspective, the 'certain weight' to which Barry referred may seem to imply a universal weight or weight range comparable to our BMI ranges, but this was not quite the case. Santorio's idea of a healthful weight range, or 'healthful Standard', differed per individual and was determined by a feeling of well-being, rather than by body weight. The term 'healthful Standard' was first introduced by John Quincy, as a translation of what Santorio referred to as the *latitudinis ponderis salubris*. Quincy described this 'healthful Standard' as:

> all that Latitude of Alteration a Body is capable of undergoing with respect to its Weight without being Distempered, is called by *Sanctorius* … a healthful Standard; the greatest Weight it is capable of, is its greatest healthful Standard, and its least, the lowest Standard and between both, its mean or middle Healthful Standard; and all these are different, at different Ages and Seasons.[39]

The 'greatest healthful Standard' therefore indicated the very maximum weight a person could be before feeling physically impaired by this weight. The upper limit (and indeed the lower limit) of what any person could weigh while still being considered perfectly healthy therefore theoretically stretched for as long as the body was free from distemper and impairment of function. How high or low that constant weight was mattered much less to health than did any signs of distemper. Within this framework, the weights of Bright and Lambert—at 44 and 52 stone, respectively—would have still been considered within their individual 'healthful Standard', as they were described as physically well and perfectly capable of exercise.

[38] Edward Barry, *A Treatise on the three different Digestion and Discharges of the Human Body and the Diseases of their Principal Organs* (London: Printed for A. Millar, 1759), 196–197.

[39] For Quincy's use and explanation of the term 'healthful Standard', see aphorism Lxiii in Medicina Statica (version 1737) on p. 81. Santorio's use of the term can be found in the 1614 Latin original, *Ars de Statica medicina*, 17.

Maintaining a near-constant weight, however, was no easy feat. The quantities insensibly perspired varied according to differences of constitution, age, countries, seasons, diet, and the rest of the six non-naturals.[40] Some of these differences were outside of anybody's control, and so to offset these and other fluctuations, those who monitored their weight had to constantly counterbalance *excreta* by adjusting *ingesta* and vice versa. The first step was to weigh, monitor, and adjust the quantity of food taken in, for '[i]f it be known exactly how much Food is convenient for every Day, such a Person may easily preserve his Health'.[41] Hence Cheyne advised people '[t]o try, by a Sanctorian Chair, the Quantity and Quality of those Foods that perspire the most and soonest, and reduce the Body to pretty near always the *same* Weight'.[42]

Exercise was another habit which had to be taken into consideration in order to maintain that perfect balance. Late eighteenth-century French physician Francis de Valagin (1724–1805) echoed Santorio's argument that '[b]y Exercise Bodies are rendered lighter' when he explained:

> A proper medium should be kept up between motion and rest...we know that this medium is kept up in our bodies, if they continue in a good state of health for several years, and of *the same weight*. For if their weight increases, with a sense of dullness and heaviness, our exercise is too little, and our rest too considerable. But if their weight decreases, our exercise is too great and our rest too little.[43]

We know of at least one example of an individual who monitored his exact weight loss as a result of exercise from Samuel Pepys's diary. Pepys recorded on 2 September 1667 that the king, Charles II, desired to be weighed after a game of tennis, as it was his 'curiosity' to know how much weight he lost over a game of tennis.[44]

It was not only daily life that affected the balance of *ingesta* and *excreta*. Body weight fluctuated on a monthly basis too. This was thought to be

[40] Ibid., 47.

[41] Ibid., 186.

[42] George Cheyne, *An essay on Regimen*, (London: Dan. Browne, 1753), liii. Emphasis my own.

[43] Santorio, *Medicina Statica* (version 1737), 257; Francis de Valagin, *A Treatise on Diet, or the management of Human Life* (London: J. and W. Oliver, 1768), 228. Emphasis my own.

[44] Samuel Pepys, 2 September 1667, in *The Diary of Samuel Pepys*, Vol. 8: 1667. Edited by Robert Latham and William Matthews. London: HarperCollins, 2000, p. 419.

especially true for women, as it was believed that women's health was largely governed by their menses. Unlike men's bodies, women's bodies were believed to excrete the majority of their waste matter on a monthly basis through their menstrual blood, instead of through insensible perspiration. The reason for this can be found in gendered assumptions regarding male and female physiology. Women's bodies were believed to be weaker than men's bodies, so that their fibres did not have enough power to propel *excreta* to the pores of the skin to excrete them through insensible perspiration.[45] Instead, their waste matter collected in the body until their sheer volume occasioned a critical discharge: this was why women menstruated.[46] Menstruation was therefore considered imperative to women's health, and consequently physicians considered 'suppression of the menses' one of the most dangerous conditions from which women could suffer.[47] This physiological difference also helps explain why women's body weight was seldom discussed in the medical literature. As Santorio, Quincy, Keill, and many others focussed on insensible perspiration as the most important factor in managing body weight, women's bodies were effectively excluded from these discussions.

The only occasion when Santorio made any mention of women was when he expressed his belief that some men had to discharge in a similar fashion to women: 'Even those men who are in a perfect state of health ... once a month increase beyond their usual weight, to the quantity of one or two pounds, and at the month's end, return again to their usual standard, in the same manner as women do; but then by a critical discharge of urine'.[48] Only in texts specifically treating menstruation would readers have learned that women's plethora and thus their weight gradually increased over the course of a month and that once the menstrual flow had passed, they would return to their 'usual' weight.[49] While women's bodies were rarely explicitly mentioned in discussions about weight management, it seems that it was generally understood that both men's and women's

[45] John Freind, *Emmenologia: written, in Latin, by the late learned Dr. Freind, translated into English by Thomas Dale.* 2nd edition (London: Printed for T. Cox, 1752), 17–18, Eighteenth Century Collections Online.

[46] Ibid.

[47] Sylvia de Renzi, 'Women and Medicine', in Peter Elmer, ed, *The Healing Arts: Health, Disease and Society in Europe, 1500–1800* (Manchester: Manchester University Press, 2004), 203.

[48] Santorio, *Medicina Statica* (version 1737), 82.

[49] Freind *Emmenologia*, 45.

body weight fluctuated over the course of the month, and that these fluctuations had to be taken into account. The purpose of weighing in this period was therefore to monitor the exact *ingesta* and *excreta* and to respond to any changes in the balance between the two. If a person had found a comfortable body weight that could be maintained that person was within the 'healthful Standard'. While the 'healthful Standard' was an 'ideal' weight range, it is important to note that this was a purely personal weight range. An individual's weight was compared to the weight of the same body at various times of the day, month, and year, rather than to the weight of other bodies. This remained true throughout the eighteenth century, the only exception to this found in the 1780 correspondence between Jane Webster and William Cullen, whose discussion about the appropriate weight for a woman presents us with the earliest known instance of the concept of a weight range being applied to a demographic (in this case, women of 'middling size'), rather than to an individual. This letter is unique in that respect, and I have found no similar discussion in any other source dating from the eighteenth century. For the majority of the eighteenth century, weight ranges were used for personal reference and were not measured against a medically standardised system such as the one that we have today.

Body Weighing Practices

Because the minutiae of the body's fluctuations were easily affected by the above-mentioned factors, physicians were almost unanimous in their advice that the practice of body weighing was not a suitable method for health maintenance outside of the context of the carefully controlled medical environment. Apart from Cheyne, physicians who had partaken in weighing experiments explicitly warned that for individuals who were not medical professionals, the practice of weighing oneself was tedious and unnecessary, and perhaps even dangerous. John Quincy, who had translated Santorio's *Medicina statica* into English, warned readers of the text against attempting to follow Santorio's lead:

> I am not at all unaware how severe some will be hereupon, in requiring how often they must weigh themselves, and whether they ought to drink and eat by the Ounce? To whom I have only this to say, that Sanctorius by the Balance, has already done enough to convince any serious person of the natural Discharges, and their Proportion to one another … from which …

any Person may soon be a Judge of the present State of his Constitution, without going into a Pair of Scales.[50]

Anticipating obsessive behaviour, Quincy admitted to having deliberately omitted the 'Exactness of the Sanctorian Calculations', in his translation of the *Medicina statica,* because, for all intents and purposes, it was sufficient for readers to understand that the weight of the body was affected by eating, drinking, sleep, and exercise and that a healthful standard could be maintained through being consistent in how much one ate and exercised.[51] The weighing experiment, as it was executed by Santorio, as well as by eighteenth-century physicians such as James Keill and Bryan Robinson, was not intended to be adopted by the non-medical public as a means to measure and manage health. Ultimately, the physicians who partook in such experiment sought to establish the intricacies of what we now understand as 'metabolism'. For Santorio in particular, body weighing was but one aspect of his wider efforts to understand and achieve homeostasis through measuring not just weight, but also the body's temperature, pulse, and even external factors such as air humidity and pressure, which were believed to influence the internal processes of the body.[52] The aim of the weighing experiment was to establish certain rules about the body's processes, rather than to prescribe body weighing as a method of health management.

In the first half of the eighteenth century, discussions about and instances of body weighing remained largely restricted to medical professionals.[53] Early eighteenth-century weightwatchers constructed aphorisms based on the results of their experiments which were designed to advise readers on how to avoid fluctuations in their *ingesta* and *excreta* and maintain a nearly constant state of balance between the two. These aphorisms offered general rules of diet and exercise generally applicable to most adults who were of good health and average stature, because they allowed

[50] Quincy, *Medicina statica* (Version 1737), vi.

[51] Ibid.

[52] See 'introduction', in Barry and Bigotti, Santorio Santori and the Emergence of Quantified Medicine, esp. p. 38.

[53] There are only a few recorded instances of body weighing before 1750 taking place outside of medical contexts. One is King Charles weighing himself after a game of tennis (discussed above). Georges Vigarello mentions a request made to lieutenant general of the Paris police in 1725 to have a weighing chair installed in public places, but from the lieutenant's reply it appears that this request was denied. Vigarello, *The Metamorphoses of Fat*, 81.

readers to apply the general principle of balancing *ingesta* and *excreta* to their own personal circumstances. They did not, however, attempt any advice on what constituted a healthy, or average, weight range. Eighteenth-century physicians largely echoed Santorio's idea of a 'healthful standard', recommending to readers of their aphorisms that they maintain a state of body under which they felt healthful. Should readers fall below or above what for them personally was a healthy weight range, they were armed with sufficient advice on how to adjust their *ingesta* and *excreta* to restore the equilibrium. As far as most physicians were concerned, there was no reason why non-medical people should partake in the practice of body weighing.

Despite early eighteenth-century physicians' warnings, weighing as a method for measuring the body and monitoring health spread to the wider public. By the second half of the century, body weighing was no longer limited to physicians. There are several indications that the practice of weighing oneself became increasingly popular over the course of the eighteenth century. In England, we know that a weighing machine was made available to the public in the 1780s, by Belgian inventor John Joseph Merlin (1735–1803) as part of an exhibition of his mechanical inventions at Hanover Square in Westminster, London.[54] For the price of one shilling, visitors could be weighed using his machine.[55] Another place where members of the public could be weighed was at a grocery shop at the 'sign of the Coffee Mill', which is still hanging today above what is now Berry Bros. and Rudd, a wine merchant at No. 3 St. James Street, in central London.[56] Industrial scales originally designed to weigh coffee were used at the Coffee Mill to weigh visitors. The weights of these visitors were recorded in ledgers, a custom which started in 1765 and continued well into the twentieth century.[57] The scales had been in the coffee house since the 1740s, and so it is possible that visitors had been weighing themselves there before they began recording their weights in the ledgers. The lack of further documentation or machines relating to public weighing should not lead to the

[54] The exhibition was advertised in a pamphlet: John Joseph Merlin, *Morning and Evening Amusements at Merlin's Mechanical Museum, No. 11 Princess Street, Hannover Square* (London: 1787), https://archive.org/details/b30375642/page/6, 7.

[55] This price is advertised in: *Morning Herald* (London, England), Friday, 25 April 1788, Issue 2343, Seventeenth–Eighteenth Century Burney Collection Newspapers.

[56] I will refer to this location as 'at the Sign of the Coffee Mill' throughout this section.

[57] I am very grateful to the staff at Berry Bros. and Rudd for allowing me access to the ledgers, which are still in their possession today, at No. 3 St. James Street.

assumption that it was not practised, as there is the possibility that industrial weighing machines were used to weigh people elsewhere too.

Weighing machines designed for personal use were increasingly available from the 1780s onwards. Some newspaper advertisements placed by individuals listing their furniture for sale include weighing mechanisms such as 'Merlin's Weighing Machine' or a 'Sanctorius Balance'.[58] These may have been purchases following visits to John Joseph Merlin's late 1780s exhibition, or after seeing advertisements for his machine.[59] Further evidence of the use of personal weighing comes from London gentleman William Cole, for example, who included in his diary the following entry: '1759. Weighing myself at my Lord Montfort's House, as I often used to do, with others, there being scales for that purpose in the colonnade under the house'.[60] Pat Rogers found that Samuel Boswell weighed himself in Lord Pembroke's house in 1776 and similarly found evidence of body weighing in the diaries of Edmond Malone, Lord Palmerston, and the Duchess of Northumberland.[61] The fact that these weighing mechanisms were generally on private property might explain why further records of people weighing themselves have not been found by historians. As Ginnie Smith pointed out, individual health care often is, and was, a manner of routine private maintenance and therefore often invisible to historians when not recorded or such papers not preserved.[62] The number of people who weighed themselves might therefore be higher still than the extant evidence suggests.

Our most extensive evidence of weighing practices in eighteenth-century Britain comes from the ledgers from the Coffee Mill.[63] The grocery

[58] *Daily Advertiser*, Issue 21038, Friday, May 6 (London, 1796); *Morning Post* Issue 6445, Tuesday, December 17 (London: 1793); *Morning Post and Fashionable World* Issue 7546, Saturday, May 7, (London: 1796); All from Seventeenth–Eighteenth Century Burney Collection Newspapers.

[59] Such as the aforementioned pamphlet, *Morning and Evening Amusements at Merlin's Mechanical Museum*.

[60] Extract from William Cole's Diary, from John Nichols, *Literary Anecdotes of the Eighteenth Century*, Vol. 1 (London: Nichols, Son, and Bentley, 1812), 698.

[61] Rogers, 'Fat is a Fictional Issue', 27.

[62] Ginnie Smith, 'Prescribing the rules of health: Self-help and advice in the late eighteenth century', in Roy Porter, ed, *Patients and Practitioners: Lay perceptions of medicine in pre-industrial society* (Cambridge: Cambridge University Press, 1985), 294.

[63] As advertised on a trading card dating from 1757. The trading card is in the possession of the Museum of London, ID: Z1704/38. https://collections.museumoflondon.org.uk/online/object/120645.html.

company was founded by a widow now only known as the Widow Bourne in 1698.[64] Following the fashions of the time, the Widow Bourne focussed on coffee and teas, supplying the local coffee houses.[65] It is for this reason that the shop chose the image of a coffee mill as a sign for their business, and today, Berry Bros. & Rudd still trades under the same 'Sign of the Coffee Mill'. By the middle of the century, when the owners started to record the weight of their customers as well as their coffee, the store was run by William Pickering, the Widow Bourne's great-grandson, and his business partner, John Clarke.[66] The grocer's location and specialisation in luxury goods attracted wealthy and upper-class clientele, which is reflected in the ledgers, which contain entries of Dukes, Viscounts, and notable names including Lord Byron, Beau Brummel, and William Pitt the Younger, who was the prime minister at the time his weight was last recorded, in 1786.[67]

The earliest record in the ledgers is from 1765, when the company was run by William Pickering and John Clarke. However, the scales which were used to weigh the visitors had already been present at the store for the purpose of weighing their tea and coffee, so it is possible that William and John had started to weigh their customers before they started to record their customers' weights in the ledgers. There are a total of nine leather-bound ledgers (which are still held at Berry Bros. and Rudd at No. 3 St. James Street today), one of which is solely dedicated to the weight recordings of the women who visited the store.[68] Older pictures of the ledgers show that the books were initially organised alphabetically, with each individual ledger covering an alphabetical range.[69] It appears that the ledgers were at some point in time re-bound and re-organised into eight ledgers organised by date.[70]

[64] An overview of the history of the business, now Berry Bros. & Rudd, is available on the website of Berry Bros. & Rudd: https://www.bbr.com/about/history.

[65] Ibid.

[66] Ibid.

[67] Douglas Ajram, Charlie Harris, and Christopher McKenna, 'Berry Brothers & Rudd: Family, Reputation, and the Cultivation if the European Wine Trade', *The Global History of Capitalism Project: case* study 6, September 2018, 1–10. https://globalcapitalism.history.ox.ac.uk/files/case06-berrybrothersandruddpdf.

[68] The ledger is titled 'Ladies A-Z' and is the smallest of the ledgers.

[69] The website of Berry Bros. & Rudd pictures the ledgers as they are in their current form, organised by date range. There is an imagine of the ledgers in their earlier, alphabetised form, in Ajram, Harris, and McKenna's 'Berry Brothers & Rudd', p. 2.

[70] In their current form, each ledger roughly covers a time period, through within each ledger the entries are still organised somewhat alphabetically. The time periods the ledgers cover are not exact however. For example, while the cover of ledger 8 indicates that it con-

The total number of entries recorded in all nine ledgers is too numerous to count and certainly run into the many thousands. Here, I will focus primarily on the entries recorded in ledger seven, which is the ledger which contains the oldest records, starting from 1765. The records from this ledger run all the way up to 1800, and there are a few records which date from after 1800.[71] There appear to be records relating to a total of 916 individuals in ledger seven.[72] The entries in ledger seven suggest a relatively constant rise in weightwatchers since the records started until the year 1784, which is the year from which the records of ledger eight start. Between the years 1765 and 1784, the number of new entries in ledger seven rose from only one in 1765 to 92 in the year 1783, which is the year with the highest number of new entries in this ledger.[73]

The question as to why the visitors of the Coffee Mill came to see body weighing as a practice worth doing frequently and consistently is difficult to answer. In the medical circles, body weighing was done for the purpose of establishing rules about what we would now call 'metabolism', as we discussed above. In the ledgers themselves no reason is given for the practice of body weighing, nor for the recording of the weight of the visitors. In fact, the only recorded data is the weight of the weightwatcher, the clothes he was wearing, and the date on which the weight had been recorded. However, we can glean from these records the significance people attached to knowing their body weight. The fact that the weight was measured precisely to the half pound suggests that precision mattered. The notes on the clothing of the weightwatcher were presumably included to ensure that the added weight of clothing was consistent across all instances of weighing. Indeed, most entries started with a description of

tains records starting from 1785, the preceding ledger seven does still contain records from after 1785. The ladies' ledger is still organised alphabetically in 2023.

[71] The number of entries in ledger seven are predominantly from 1774 to 1784, after that date range the entries become more sporadic. The ledger has almost no entries for 1785–1789 (most of which are recorded in ledger 8), but then records 86 entries for the year 1790. Ledger seven contains 27 records from post-1800, the latest record dating from 1990.

[72] This ledger contains the oldest records and due to their age, some entries are difficult to read. In addition, the records have been re-organised over time and it is possible that records of some individuals have been counted twice or have been missed by myself. There are 10 records which I have not counted towards the 916 due to them being illegible or undated.

[73] It is important to note that with new entries, I mean new individuals who started to weigh themselves in that year. There were more than 92 individuals who weighed themselves in the year 1783 if we were to also count all the individuals who had started to weigh themselves in prior years and continued to do so in 1783.

clothes worn on the first instance of weighing ('regimentals', 'shoes', and 'boots' were frequently noted down), after which most entries recorded 'ditto' to indicate the same type of garment was worn again on subsequent instances. Some weightwatchers seemed to have taken at least some of their garments off, presumably to obtain a 'truer' value. On 10 April 1788, Mr. Braddyk was weighed first with boots, and then again 'stript'.[74] It is possible that Mr. Braddyk wanted to know exactly how much weight his boots added so that he could subtract this from the total weight on future occasions. The weighing device was (and indeed still is) in the grocery store itself, close to the store entrance, which would have offered little privacy. It is possible that the owners may have closed the store to allow privacy for some of their clientele. Whatever the case may have been, these notes on the garments of the visitors indicate that weightwatchers wanted to ensure that the weight recorded accurately reflected the weight of their bodies and not of their clothing.

The recording of the dates on which people weighed themselves can also be indicative of the purpose behind the practice. We see not only when people started to weigh themselves, but also the frequency with which they did so. In the first years of recording, many visitors of the Coffee Mill weighed themselves only on one occasion, while visitors in later years often came back to track their weight over longer periods of time and weighed themselves with greater frequency. In the year 1768, for example, five out of the total of ten visitors only weighed themselves once, four visitors weighed themselves two to five times per year over a period spanning less than five years, and only one individual traced his weight for a period of longer than five years. In the year 1783 we find that just under a third of the 92 new visitors weighed themselves only once, while more than half of visitors who started weighing themselves in that year would continue to keep track of their weight for a period of more than five years. Fewer than half of the 92 new entries in 1783 show people weighing themselves once in that year, with more individuals choosing to weigh themselves two to four times a year. A slowly growing minority (11 individuals in 1783) weighed themselves more than once a month at the Coffee Mill.

In her analysis of seventeenth- and eighteenth-century weighing practices, Dacome argues that physicians had managed to convey the more general idea that the internal balance and health of the body could be

[74] Entry for Mr. Braddyk. Ledger 7, f. 20, Berry Bros. and Rudd.

found and scrutinized by weighing, and the data from the Coffee Mill ledgers support such a conclusion.[75] Clearly weighing oneself at the Coffee Mill became an increasingly popular practice between 1765 and 1784, as we can see from not only the increase in the number of persons who had started to weigh themselves, but also in the way they were partaking in the practice. While we could surmise that perhaps the earliest weightwatchers weighed themselves out of curiosity and without much further interest in body weighing as a practice of health management, for many entries from the 1780s such an explanation will not suffice. The Earl of Cholmondeley, for example, weighed himself at least 37 times over a period of five years, usually monthly, but sometimes more often.[76] These instances of relatively frequent and long-sustained participation in body weighing practices suggest that weightwatchers attached value to the practice of weighing oneself. What is more, entries such as these suggest that many individuals were not just recording, but also monitoring their weight, similarly to Santorio and later physicians such as Robinson and Keill, who recorded and monitored their weight over the course of the eighteenth century.

As discussed above, the physicians who had partaken in body weighing experiments monitored minute shifts in the weight of their bodies in order to ascertain, and to some extent control, the effects of external factors such as food and drink, exercise and sleep, and even temperature, on the weight of their bodies. In these experiments, body weight was seen as indicative of the body being in a state of healthful equilibrium (in the case of a constant weight) or not (in the case of weight loss or weight gain). The way in which many of the visitors of the Coffee Mill weighed themselves suggests that even outside of the medical circles body weighing was practised for the purpose of monitoring shifts in weight frequently and over long periods of time. These weighing practices make sense when we consider that Santorio had argued that an unchanging body weight was a sign of good health, and a changing body weight, a sign of ill health.[77] Repeated body weighing was one method in which people could gauge

[75] Dacome, 'Living with the chair', 469.

[76] Entry for The Right Honourable Earl of Cholmondeley, Ledger 7, f. 24, Berry Bros. and Rudd.

[77] Santorio, *Medicina statica* (version 1720), 51. Santorio stated in his 15th aphorism, 'If a Body returns to the same Standard every Day, without any Change in the Quantity of Perspiration, there will be constantly preserved a perfect Health'. In his 16th Aphorism, Santorio added: 'Bad Qualities arise when the Body is not one Day the same in Weight as another'.

whether their weight was changing or unchanging and, therefore, whether their state of health was changing or unchanging.

However, there are some considerable differences in the weighing practices recorded in the ledgers, and the weighing practices of physicians such as Santorio, Keill, and Robinson. The ledgers demonstrate that while some people did weigh themselves regularly (some did so on a monthly basis, many others once per year), there are no records of any visitors who weighed themselves daily. This discrepancy is unsurprising, but relevant. Santorio and the later experimenters monitored shifts in the body's weight within a 24-hour period in particular. To know exactly how much the body lost or gained daily, experimenters weighed themselves not once but many times throughout a 24-hour period, noting down every shift in weight after urinating, or consuming a meal, for example. Without such frequent weighing, any conclusions about what caused a shift in body weight was guesswork, because there were so many different factors which could affect the body's weight. Everything from exertion, temperature, humidity, changes in seasons, and even particular food items were known to bring about shifts in body weight. Inconsistent body weighing practices, in an uncontrolled environment, physicians warned, could lead to errors and misinterpretations.[78]

Despite these warnings, a less precise approach to the practice of body weighing clearly did develop at the Coffee Mill. This difference between medical and non-medical weighing practices appears to primarily be a matter of practicality. Not many individuals would have had access to personal weighing machines at this time, and daily visits to the weighing chair at the grocery shop might not have been convenient for most people. Moreover, such a deviation from the precision-focussed weighing experiment to more practical forms of monitoring and managing one's health was recommended by several physicians at the time. When it came to dietary quantities, for example, physician George Cheyne recommended that people first determine 'by Weight or Measure' how much daily food they needed, after which people were free to estimate 'by our Eye'.[79] It is possible that, similarly to Cheyne's recommendation, visitors of the Coffee Mill weighed themselves occasionally and tried afterwards to gauge (perhaps with reference to the fit of their clothing, or visual perceptions of

[78] Quincy, *Medicina Statica* (1737 version), iv and 203–204.

[79] George Cheyne, *An Essay of health and long life* (London: Printed for George Strahan), 40.

changes to their bodies) whether any significant change in weight had taken place. Weighing yearly may have been considered sufficient for the purpose of maintaining a relatively constant body weight if the emphasis on precision in measuring had started to wane.

There are some other factors which may explain the discrepancies between medical and non-medical weight practices. The Coffee Mill was a public, social setting rather than a private one, and so we can infer that the context of body weighing was one of sociability, as well as routine health maintenance. This need not imply that weighing was taken less seriously, but rather that body weighing at the Coffee Mill was a shared experience between people of relatively diverse, non-medical backgrounds, which could have led to a weighing culture which differed from the medical approach. Within such a culture, it is possible that the weighing practices of one visitor were informed by the weighing practices of other visitors, rather than by instruction found in medical literature.

At first glance, there appear to be no women in the Coffee Mill ledgers for the period under consideration. Berry Bros. & Rudd hold two ledgers titled 'Ladies', but most of the entries in these ledgers are dated after 1885. As Dacome pointed out, such an absence of women may have been due to various reasons from concerns around decency to ideas about women's bodies which made weighing them redundant.[80] Another hypothesis to consider is that women did not frequent the places where public scales were kept. The Coffee Mill may have been an establishment more usually frequented by men than women. Whether this would also apply to John Joseph Merlin's exhibition is less certain. Merlin did invite both ladies and gentlemen in his advertisement: whether they weighed themselves there is not known.[81] That being said, there is at least one entry in ledger 7 (dated 1784) that appears to be for The Honourable Mrs. Lisle.[82] Mrs. Lisle's entry is separated from the entries for the men by a large open space, not given a proper bracket, and details for her dress are not included, whereas all other entries have at least some comment on dress. In ledgers 8 (1785 onwards) and 6 (1790 onwards), we find another 34 entries for women at the very back of these books. The impression is therefore that these women's entries were unusual. Most of the women recorded in these ledgers weighed themselves only very occasionally, with

[80] Dacome, 'Living with the chair', 489.

[81] *Morning and Evening Amusements at Merlin's Mechanical Museum*, front page.

[82] Entry for Mrs. Lisle, ledger 7, fl. 29, Berry Bros. and Rudd.

27 out of 34 weighing themselves only once.[83] The Honourable Mrs. North weighed herself in 1786 and in 1787, two to three times each year.[84] On most of the dates she was weighed, Mr. North was weighed also. This suggests that women were weighed only occasionally, and usually when in the company of their husband (or a male relative). This, in combination with the manner in which women's weight was recorded in the ledgers, is suggestive of a weighing culture in which women did not usually participate but did so on occasion.

There was, however, at least one woman who did weigh herself with great interest and some regularity. The letters written by Jane Webster, the physician's daughter who had contacted Cullen in the hope of obtaining a 'moderate' weight of 12 pounds, indicate that even if body weighing had not yet become commonplace among women or even most men, the connection between body weight and corpulency was now explicit in medical circles. It is not clear why this shift might have taken place towards the end of the eighteenth century. The increased availability of weighing machines, both personal and in public settings such as the Coffee Mill, may have contributed to a weighing culture in which access to information regarding body weight was more commonplace than it had ever been, allowing for it to be shared and compared, consequently leading to ideas of what constituted a 'moderate' weight for women and for men. However, as Cullen and Webster both had medical backgrounds, the evidence currently points to a shift in medical theory regarding body weight and corpulency. This seems all the more likely if we consider that this epistolary conversation concerns weight ranges for women, when women's participation in popular weighing practices was, by all accounts, much lower than men's participation.

There is another letter addressed to Cullen in which body weight is discussed in relation to corpulency, dating from 1773.[85] In this letter, the patient described himself as having been 18.5 stone in weight and five feet eight inches tall in height, connecting weight and height in a manner similar to Webster, but rarely before seen in discussions of corpulency.

[83] The entries for the women mentioned here can be found in ledgers 8, 7, and 6. Other entries for women weighing themselves at the Sign of the Coffee Mill can be found in a separate 'Ladies' Ledger.

[84] Entry for Mrs. North, ledger 8, f. 34, Berry Bros. and Rudd.

[85] Case note regarding: Anonymous (patient), 26 April 1773 (Doc. ID:844), 3, Consultation Letters of Dr William Cullen (1710–1790) at the Royal College of Physicians at Edinburg, accessible online via The Cullen Project, http://www.cullenproject.ac.uk.

Furthermore, when this patient speaks of having reduced himself, he seems to do so with reference to his weight, rather than his bulk or plethora. Both factors point to weight becoming part of the discussion of health between physician and patient in a way that it had not been before. It is possible that a wider availability of weighing machines provided patients and physicians alike with the tools effectively to communicate matters of body size, type, and health. Patient-physician correspondence was often very detailed, and this new form of measurement may have been a helpful addition to provide the physician with the fullest possible picture. While this explains body weight appearing in correspondence in the late eighteenth century, the inclusion of body weight does not necessarily mean that it was paid much attention or ascribed much value in diagnosing or curing corpulency. Despite him still weighing 17 stone at his time of writing, this patient described himself as active and in good health, and his letter contains no indication that he considered his weight to be *too* high—unlike Webster, he did not ask Cullen for any advice for further reducing this weight, and unfortunately, we have no record of Cullen's response. While on the one hand this letter does point to weight becoming more commonplace in late eighteenth-century discussions about fatness, it is also clear that this patient hailed from a time still very far removed from our modern ideas of what constitutes a healthy weight, and a healthy body.

Conclusion

In this chapter I have argued that in order for us to understand eighteenth-century body weighing practices, we need to first challenge our assumption that body weighing practices and concerns about fatness were always intrinsically connected. Corpulency was a topic that was increasingly written about over the course of the seventeenth and eighteenth centuries, and fat bodies were increasingly medicalised and stigmatised. Some historians have understandably considered the eighteenth century as the beginning of modern concerns about obesity, and, consequently, modern weight-watching practices. We have seen, however, that for the majority of the eighteenth century, body weight did not yet play a role in the medicalisation of corpulency. While there are signs that this started to change by the close of the eighteenth century, body weight did not generally feature in eighteenth-century understandings of fatness or corpulency.

The use of scales to weigh the human body in a medical context was first introduced by Santorio in the early seventeenth century, but his weighing experiment was later repeated by physicians over the course of the eighteenth century, and the practice of body weighing was adopted outside the medical circles as well, increasingly so from the second half of the eighteenth century onwards. From the contemporary medical literature on the topic of body weight, we can clearly discern that the purpose behind eighteenth-century weighing experiments undertaken by physicians was the pursuit of a healthful balance between *ingesta* and *excreta*, which, according to Santorio, an unchanging bodyweight was evidence of. How much or how little one weighed mattered insofar as it informed the physician of body weight changes after eating, drinking, perspiring, and excreting—there was no concern yet for how much the average person weighed, and there was not yet any concept of 'overweight' or 'underweight' in the way that we use those terms today. As we have seen in the cases of Daniel Lambert and Mr. Bright, even the heaviest bodies could be considered healthful so long as people felt healthful and their weight fluctuated but little. While there are some indications that towards the end of the eighteenth century a heavy weight started to become seen as a medical problem in itself, for much of the eighteenth century this was not yet the case.

The general Santorian idea that body weight could inform people of the state of their health clearly survived into the eighteenth century, as did the notion that weighing was best done with some regularity and precision. However, the minute and regimented nature of Santorio's experiment seems to have mostly disappeared from eighteenth-century practices among non-medical weightwatchers. As the weighing ledgers from the Coffee Mill show us, many non-medical people weighed themselves yearly, or even monthly, but not every day. They made the effort to weigh themselves frequently and for many years and took care to get a 'true' measurement of their weight by taking the weight of their garments into consideration. They did not weigh themselves after every meal and every evacuation, however. The ledgers therefore are a clear indication that the practice of body weighing had deviated from the experiments undertaken by Santorio in the early seventeenth century, and those undertaken by physicians in the early eighteenth century.

I have suggested that a combination of physicians' warnings against 'precision measuring' with practical consideration and perhaps even local customs are the primary causes behind the deviation from the seventeenth- and early eighteenth-century weighing experiments and popular weighing practices in the late eighteenth century. First and foremost, there was the matter of practicality. Physicians had warned readers of their texts that Santorian-esque weighing experiments were not feasible for the average person. Personal weighing devices were not yet widely available in the eighteenth century, so weightwatchers were required to travel to the few weighing chairs available to the public, such as the one at the Coffee Mill in London. It is also possible that the public setting of the weighing chair at the Coffee Mill fostered a weighing culture distinct from the weighing culture of the medical profession. The weightwatchers of London may have relied on each other rather than on the reports of medical weighing experiments for guidance on how often or for what purpose they should weigh themselves. This is conjecture, of course, as the weighing ledgers contain no information as to why people weighed themselves at the Coffee Mill.

Nevertheless, these weightwatchers' dedication to the practice of body weighing indicates that body weight in the eighteenth century was a weighty matter for many people. How it was practised differed depending not only on the period under consideration, but also on the environment in which the body was being weighed, whether medical or social. This chapter has called attention to these differences by examining the medical literature on the topic of body weighing, as well as the weighing ledgers from the grocers at the Coffee Mill. Based on these findings, it is the contention of the author that eighteenth-century body weighing practices, while certainly reminiscent of Santorio's experiment, had deviated considerably from how Santorio had weighed himself. The practice of body weighing had not yet developed into a practice we might recognise as similar to our own approach to body weighing however. We must therefore be careful to present eighteenth-century body weighing practices neither as a linear development from Santorio's weighing experiment nor as the beginning of a modern practice. The practice of weighing oneself may appear universal, but, as I have shown, it can and did have different meanings to different people in different times and contexts, meanings which inform how body weighing was and is practised.

CHAPTER 7

Conclusion

As a period in which both old and new ideas thrived, the eighteenth century deserves much more attention from medical historians than it has hitherto received, especially when it comes to topics such as diet, exercise, and body weight. This study has sought to bring these three subjects to our attention and to the fact that eighteenth-century health guides were a site of considerable change, as well as continuity. Health advice in popular health guides shifted in response to new theories and research practices brought about by the iatromechanical school of thought. As each individual chapter has shown, such shifts can be very informative both of changes in medical thinking and of the relationships between medical theory and popular health advice. These developments deserve our attention even if health advice often retained some of the older frameworks and ancient medical theory and failed to result in lasting change, because historically singular theories or schools of thought can be particularly useful for understanding how and why some ideas continued while other changed. Furthermore, eighteenth-century health guides help to shed light on our modern ideas about health and the body, especially related to urbanisation, industrialisation, and new health technologies. Insight into the role of quantification and measurement in eighteenth-century health literature may help us reflect on the potential effects of our modern health

M. Wübbels, *Quantification and Measurement in British Health Advice, c. 1670–1830*, Palgrave Studies in Medieval and Early Modern Medicine,
https://doi.org/10.1007/978-3-032-32760-4_7

technologies and current culture of 'Self-Tracking'.[1] Recent studies examining the relationship between our modern concept of personhood and self-tracking have investigated the history of self-tracking in order to ascertain how this relationship developed over time, and this study contributes to that ongoing discussion by shedding light on eighteenth-century attempts to maintain health through the practices of body weighing and the quantifying of our daily dietary needs.[2]

From this study, it has become clear that health guides are also an excellent source for gauging the interactions between medical professionals and their readers. Although any research which focusses on these texts faces some challenges in terms of determining whose ideas and practices they realistically represented and how far their influence reached, they also offer a unique insight into the considerations physicians took into account in deciding how their knowledge could result in practical application by their readership. It has become evident that eighteenth-century health guides represent ideas unique to their time and medical milieu. By focusing on the interaction between early eighteenth-century iatromechanical theory and the re-purposed seventeenth-century Santorian principles, this book has demonstrated how theories about health and the body were the result of a complex mixture of fashionable ideas, traditional knowledge, new technologies, and even non-medical circumstances.

The question of the extent to which professional medicine and popular health were interlinked or remained separated remains a central one to scholarship of this field. While an in-depth analysis of the readership of these health guides is a matter which is beyond the scope of this book, I have demonstrated that popular health advice was closely connected and highly responsive to shifts in medical theory and methodology. Furthermore, it has become clear from this study that physicians constructed their health advice with not only the medical theory in mind but

[1] For discussions about our modern culture of self-tracking, see: Btihaj Ajana, Joaquim Braga, and Simone Guidi, "Introduction", in Btihaj Ajana, Joaquim Braga, and Simone Guidi (Eds.) *The Quantification of Bodies in Health: Multidisciplinary Perspectives* (Emerald Publishing Limited, Leeds 2021) pp. 1–12. https://doi-org.uoelibrary.idm.oclc.org/10.1108/978-1-80071-883-820211001; Btihaj Ajana, "Introduction: Metric Culture and the Over-examined Life", Btihaj Ajana, (Ed.) *Metric Culture* (Emerald Publishing Limited, Leeds, 2018) pp. 1–9. https://doi-org.uoelibrary.idm.oclc.org/10.1108/978-1-78743-289-520181001.

[2] See especially: Fenneke Sysling "Measurement, Self-Tracking and the History of Science: An Introduction", *History of Science,* 58, 2, (July 2020), 103–116.

taking into consideration also the readers' ability to put theory into practice. Printed health advice therefore does not solely represent the physician's voice but also gives us a glimpse of what non-medical people considered to be reasonable or unreasonable practice. Nevertheless, there is little doubt that these guides were primarily written for a fairly wealthy clientele, and so the dietary, exercise, and weighing practices of the poorer members of society remain largely unknown. Future studies may look at larger quantities of letters or other personal documents, although those kinds of sources also tend to over-represent the elite members of early modern and eighteenth-century society. A more effective approach to studying weighing practices may be to ascertain the number and locations of eighteenth-century weighing machines, especially public weighing machines such as the one displayed by Merlin in central London in the 1780s.

Chapter 2 assessed discussions about the choice of aliment and found that this continued to be an all-important factor in health maintenance. The resilience of many traditional aspects of dietary advice should not lead us to assume that this field of medicine was unaffected by the onset of iatromechanism in the early eighteenth century. There were some considerable developments inspired by ideas and theories born out of this school of thought, as physicians turned their attention to the machine body's countless vessels tasked with the circulation of the nourishing fluids. Considering the importance of the satisfactory functioning of these vessels to the body's overall health, much advice found in health guides started to focus on their maintenance. Dietary advice of the early eighteenth century reflected this concern. Firstly, the authors of health guides discussed the quality of aliment not primarily in relation to its humoural qualities but also in terms of its consistency. Food was not just hot or cold but also cohesive or frangible, digested with difficulty or ease, slowly or quickly broken down and assimilated into the body or cast out through the excretory channels. Secondly, the very purpose of food became re-defined by some physicians. Diet was no longer a matter of maintaining or obtaining balance in the complexion by the early eighteenth century. Rather, physicians advised that the purpose of food was to repair and replenish what was daily lost through excretion and exertion. I have suggested that the idea that food needed to 'make up' for such daily losses fostered a quantitative approach to diet wherein physicians measured what was lost in excretion through weighing experiments and calculated exactly the weight of food the body needed for its replenishment.

The chapter furthermore demonstrated that concerns for food qualities did not disappear but that new qualities such as 'nourishing' or 'perspirable' became increasingly discussed while older, Galenic qualities such as hot or cold featured less in these discussions. These newer qualities reflected physicians' search for diets which answered for the daily losses measured by their weighing chairs. Nourishing foods added much to the body and were appropriate for those who needed to repair and replenish much, whereas perspirable aliment was mostly excreted and therefore suited to those who did not need to make up for losses sustained by hard labour or illness. This divide between nourishing and excretory aliment seems to have led to similarly divided courses of diet. Diets consisting of mostly nourishing foods were thought suitable for the working population, while excretory or 'perspirable' foods of easy digestion were hailed as proper for those who were sedentary and for those who suffered from impaired digestion. As the sedentary, the aged, and those who were concerned about their health and diet made up the bulk of their readership, physicians such as George Cheyne and Antonio Cocchi soon promoted dietary courses which certainly appeared standardised in their promotion of perspirable or 'light' foods to all their readers. Although it is tempting to see this development as a prelude to modern diets which are fairly indiscriminatory in their promoting of weight loss regardless of constitution, sex, age, or environmental factors, it is important to recognise that the courses of diet promoted by Cheyne and Cocchi were heavily criticised for being far too general in their approach and that their works did not reflect the whole genre, which continued to approach diet in a bespoken manner until the nineteenth century. These shifts therefore do not constitute a linear transition from early modern bespoke to modern standardised diets but rather illustrate the changeable and contested nature of eighteenth-century health advice.

In Chap. 3 I showed that the right quantity of food became a topic of intense debate in the early modern period and especially so in the early eighteenth century. I assessed whether the period under consideration witnessed a shift in dietary advice from using appetite as a guide to using a measure for determining how much a person should eat daily. As we have seen, the influx of exotic delicacies such as spices that provoked the appetites and French and Italian cooking recipes for rich sauces and tantalising dressings had physicians concerned that the traditional approach of relying on sensations of appetite and fullness could no longer be trusted. Calls for more reliable methods to determine dietary quantities were answered by iatromechanical physicians who proposed to measure and calculate how much a person ought to eat on a daily basis. They were not the first to attempt this: already in the sixteenth century, Santorio had undertaken an extensive weighing experiment designed to establish how much aliment the human body needed to consume to make up for its excretions. The aphorisms resulting from this experiment were translated into English and republished by iatromechanical physicians looking to use Santorio's findings to support their new arguments. As I have shown, though, early eighteenth-century quantification methods were not the same as Santorio's. Unlike Santorio, later experimenters did not recommend that others follow their example in weighing their bodies and basing their diets around the outcome. Daily weighing was cumbersome, could lead to obsessive behaviour, and was fraught with inaccuracies if not carefully done, they argued. It was also deemed unnecessary: despite the authority many physicians ascribed to quantification, the overall consensus was that the cues of the body remained the surest guide to determining what quantity of aliment was healthful. These cues could not always be trusted and the best way to make sure sensations of hunger and fullness were accurate was to combine qualitative with quantitative methods of measuring dietary intake. As we have seen from treatises and letters, the quantitative methods in use in this period included weighing food, counting mouthfuls, and counting stools, as well as weighing the body. While this period certainly witnessed a shift towards quantitative methods for measuring diet, there was no significant shift from using qualitative to quantitative methods for managing dietary intake.

Nor did these weighing experiments cause advice regarding dietary quantities to become standardised. Eighteenth-century experimenters were aware that their bodies and living circumstances would not match with those of many of their readers and were careful to explain how the different factors affected measurements of body weight and *excreta*. In fact, the eighteenth-century experiments were undertaken because of this very premise. Eighteenth-century physicians James Keill and Bryan Robinson repeated Santorio's weighing experiment in England and Ireland so that the British might have access to aphorisms more applicable to them and their particular environmental circumstances. Both physicians agreed that the measurements undertaken by Santorio in Italy could not possibly be representative of people living in different cooler temperatures. Other factors were also understood to affect measurements such as age, constitution, and illness, and both Keill and Robinson included these in the introductions to their tables so that every discrepancy between their and other experimenters' measurements was sufficiently explained. By disclosing their circumstances these experimenters acknowledged that their personal measurements and the quantity of aliment they required were not representative of all bodies. They recommended readers glance over the tables to learn which factors affected *excreta* and then consider how these affected their personal circumstances, rather than copy whatever dietary quantity Santorio, Keill or Robinson had set for themselves. Decisions regarding dietary quantities generally continued to be made on a case-by-case basis, in which the age, constitution, illnesses, and lifestyle of an individual were taken into account. That being said, there were some clear attempts at setting quantities appropriate for large groups of people. Cheyne famously suggested 8 ounces of meat and 12 of vegetable food was enough to keep a man in health—although he limited this recommendation to men of sedentary lifestyles specifically. Other physicians were similarly bold in prescribing certain quantities to people above the age of 50. These two factors appear to have invited a certain level of generalisation in terms of dietary rules, and I have suggested that this was because physicians considered both the aged and the sedentary to suffer from the same symptom, namely impaired digestion.

As discussed in both Chaps. 2 and 3, physicians increasingly prescribed diets designed specifically to cure certain diseases, ailments, or symptoms, rather than to maintain humoural equilibrium. Factors which were so important in humoural dietary advice, such as constitution, gender, and age, were given less space in the dietary regimen prescribed for persons suffering from weak digestive powers. Gender constituted a major factor in traditional Galenic medicine where women's bodies were understood to be colder and wetter than men's, but in the eighteenth century, sex difference hardly featured in discussions about dietary quantities. In fact, Cheyne suggested that people of both genders suffered from the same digestive issues once they reached the age of 50, and Hill too argued that aged bodies shared physiological issues which were more relevant in constructing appropriate diets than any differences in constitution. The focus on shared symptoms rather than humoural differences allowed eighteenth-century physicians to prescribe dietary rules which were more widely applicable than they had been before. While a number of physicians attempted to write dietary advice applicable to large groups of the population, many other physicians continued to argue that the best dietary advice was that which carefully considered the needs of the individual.

Decisions about diet could not be made without proper regard for exercise, which was a non-natural to which eighteenth-century physicians ascribed similar and sometimes even more importance than to diet, as Chap. 4 has shown. Although exercise advice has not enjoyed the same level of attention from historians as has diet, eighteenth-century physicians rarely failed to warn of the detriment to health caused by a lack of exercise. A care for due levels of exercise was not new to the eighteenth century, but this non-natural did receive particular emphasis from iatromechanical physicians who believed it vital to the maintenance of the solid fibres of the body. Interest in the solid, rather than the fluid, part of the body started to develop in the seventeenth century and reached its peak in the early to mid-eighteenth century. According to physicians of the iatromechanical school, the solid fibres of the body played an active role in the distribution of nourishment throughout the body by both containing and propelling forth the body's fluids. The mechanists' focus on the health and functioning of the solid fibres of the body meant that exercise, as the primary method for strengthening these solids, enjoyed levels of attention it had not previously received. With the increased importance ascribed to exercise, the lack thereof quickly came to be considered one of the foremost threats to health. Physicians lamented the sedentary professions, modern transport methods, and leisurely pastimes of their elite clientele and assessed how these individuals could make up for the faults of such lifestyles. Sedentary professions and pastimes were not new to the early eighteenth century and so the cause for this increased discomfort with sedentary lifestyles must be located in shifts in medical theory rather than in significant changes in occupational or leisure activities, which thus far have not been reported by historians to have taken place in this period.

The promotion of exercise and vilifying of sedentarism became not only more prominent in health guides but the nature of exercise advice changed as well. The theory that the body consisted of millions of minute fibres led to physicians recommending activities which exercised all those fibres equally. To achieve this, physicians recommended anything which sent minute vibratory motions throughout the entire body. Horseback riding, sitting in a carriage, and walking were all particularly appropriate as they 'shook' the whole body and reached even the smallest fibres. Even listening to music or vomiting were suggested to count as salutary exercises for their ability to affect fibres in this manner. Consequently, the exercises which came to be recommended tended to be those perceived to be of a more 'passive' nature than what was recommended in traditional texts.

The variety of exercises also narrowed because of this focus on vibrations. These kinds of exercises continued to be recommended over the course of the eighteenth century but did not survive into the nineteenth century, when physicians started recommending gymnastic exercises and strength training. The minute and vibratory exercises which we find recommended in eighteenth-century health advice therefore appear to be strictly related to iatromechanical theories of the body and were an aspect unique to the health advice of this period.

Exercise was also decidedly less sweaty in the eighteenth century or so physicians argued it should be. In Chap. 5 I have shown that fears around sweating during exercise were deeply entwined with contemporary anxieties around the balance of the *ingesta* and *excreta*, and that these fears permeated the health literature of this period. In the eighteenth century, perspiration and sweat were argued to be two different things. The first was healthful, the other detrimental. Perspiration and sweat differed in form, origin, and effect upon the body and were antithetical to one another: an increase in sweat necessarily caused a decrease in perspiration. This is what Santorio had argued in his *Medicina statica*, and upon its re-publication in English in the early eighteenth century, these ideas came to be repeated and elaborated upon in the British health literature. Early eighteenth-century physicians feared that sweating might disrupt insensible perspiration and therefore recommended that readers be moderate in the exercise they took. Readers were told to monitor their perspiration levels and cease exercising before perspiration turned visible. Physically intensive exercises were best avoided while more passive exercises such as horseback riding and walking were recommended, not only because these exercised all the fibres evenly but also because these activities could be undertaken without breaking a sweat. Although this advice was almost omnipresent in health guides discussing exercise, it did not apply to everybody. Physicians remained firmly of the opinion that every single body had unique characteristics and therefore similarly unique exercise needs. Fat bodies required significant evacuations and certain physicians therefore suggested that they did exercise to a sweat. The old were already running out of moisture since, in a continuation of ideas about humours and the changing constitution during the lifecycle, they were seen as becoming drier in old age: any exercise which exacerbated that loss of moisture was even more detrimental to their health than it was for other, younger people. Just as with diet, the right quantity of exercise one needed was relative to one's *ingesta* and *excreta* and one rule could not possibly apply to all.

Nevertheless, what *was* universal was the rule that everyone should pay close attention to their sweat and perspiration and adjust their exercises accordingly.

The idea that exercise should be 'ad ruborem, non sudorem' (to redness, not to sweat) was repeated by physicians throughout the eighteenth century, which indicates that this aspect of Santorian health advice had more longevity than some of his other theories. I have argued that this can be explained by the relative ease with which this particular theory was incorporated into vitalism, a medical school of thought popular in the late eighteenth century. Vitalists shared the mechanics' concerns with the body running out of vital matter through what they considered to be an excessive evacuation of fluids which still contained some nourishment. The Vitalists' interpretation of what exactly the body would run out of and how it did so resembled classical theory more than it did mechanical theory, but following some adjustments in terms and explanations, the general message that exercise ought to stay sweat-free remained largely unchanged until the early nineteenth century.

Concerns with the careful balancing of the *ingesta* and *excreta* of the body are perhaps most clearly reflected in the onset of a new health practice in the eighteenth century: body weighing. Although Santorio had started to weigh himself in the sixteenth century, this practice became more widely practised in the mid-eighteenth century, both by medical professionals and non-medical people. This development has been described by some as the beginning of our modern weight-watching practices. However, the last chapter of this book has demonstrated that eighteenth-century body weighing practices were shaped by the medical theory of their time and had very little in common with how we use the scales today. Generally unrelated to concerns about fatness, eighteenth-century weighing practices were informed by the Santorian theory that to maintain health, the *ingesta* and *excreta* of the body ought to be in balance. Weighing the body, Santorio suggested, could help monitor any shifts in quantity between what was consumed and what was excreted. Medical discussions about body weighing focussed on this theory and the evidence of instances where people weighed themselves in this period suggest that they mostly did so motivated by these Santorian principles. Corpulency did not generally feature in these discussions, and body weighing did not usually feature in discussions about corpulency. I have therefore argued that these two topics, now so intricately connected, remained almost wholly separate for the majority of the eighteenth century, which again underlines the singularity of the medical milieu of this period.

The idea that body weight could inform people of the state of their health clearly spread from the strictly medical circles into the mindset of non-medical people as well. However, popular weighing practices outside differed from the experiments undertaken by Santorio, Keill, and Robinson. As the weighing ledgers from the grocers at the Sign of the Coffee Mill show us, those non-medical people who weighed themselves did so yearly, or even monthly, but not every day. They made the effort to weigh themselves frequently and for many years and took care to get a 'true' measurement of their weight by taking the weight of their garments into consideration. They did not weigh themselves after every meal and every evacuation, however. The ledgers therefore are a clear indication that the practice of body weighing had deviated from the experiments undertaken by Santorio in the early seventeenth century and those undertaken by physicians in the early eighteenth century. I have suggested that a combination of physicians' warnings against 'precision measuring' with practical consideration and perhaps even local customs are the primary causes behind the deviation from the seventeenth- and early eighteenth-century weighing experiments and popular weighing practices in the late eighteenth century.

Concerns with beauty and body shape therefore were not as prominent in medical discussions of the body and health as we may have expected of a society which was otherwise very preoccupied with a beautiful, 'polite' body. Despite the increasing emphasis on beauty and exercising control over the external body, eighteenth-century health discussions continued to emphasise the importance of the internal, rather than external, when it came to determining whether a body was considered healthful or not. As Chaps. 3 and 6 have shown, increasing occurrence of body weighing practices should not be interpreted as a consequence of new beauty standards and pressures to lose weight but rather as part of the contemporary medical culture which favoured measurement to confirm the well-functioning of the digestive system. Weighing machines were not used to combat fatness or exercise control over the shape of the body but to warn the weight-watcher of any alterations in the physiological processes of the body, which could indicate illness. This discussion provides necessary context to ongoing discussions in the history of the body, as it illustrates the ongoing importance authors and readers of health guides ascribed to the internal workings of the body in their imagining of the 'ideal' body.

Taken together, these chapters illustrate the complex and contested nature of health advice in this period, an appreciation of which may lead to more future scholarship on this topic. This study might encourage more interest in, especially, iatromechanism and the effect of this school of thought on health advice, which sometimes finds itself mentioned briefly in between more extensive discussions of Galenism and iatrochemistry. This appears to be especially true for studies into seventeenth- and eighteenth-century dietary advice, many of which focus on the shift between humoural and chemical interpretations of aliment but forgo analysis of how mechanist physicians imagined the role of food in maintaining the body and how this affected dietary advice. This study has shown that it was physicians of the mechanical school of thought especially, who appropriated the principles of Santorio and applied his quantitative methodologies to dietary and exercise advice. Similarly, mechanically inclined physicians played considerable roles in the outcry about sedentarism, concerns with perspiration, and nascent weighing practices. This school of thought evidently had a significant effect on contemporary health advice, and any future inquiry into the health beliefs and practices of this period must carefully consider all the theories that informed them.

Bibliography

Primary Sources

I: Online Sources

Addison, Joseph. 1711a. No. 25. *The Spectator*. Edited by Addison, Joseph, and Steele, Richard. 1, 25, March 29, 92–95. https://archive.org/details/in.ernet.dli.2015.39152/page/n101

Addison, Joseph. 1711b. No.115. *The Spectator*. Edited by Joseph Addison and Richard Steele, 1, 115, July 12, 120–23. https://archive.org/details/in.ernet.dli.2015.39152/page/n437/mode/2up?q=dumb+bell.

Addison, Joseph. 1711c. No. 195. *The Spectator*, 3, 195, October 13, 98–99. https://archive.org/details/in.ernet.dli.2015.502846/page/n107/mode/2up.

Anon. 1818. *The Life of that wonderful and extraordinarily heavy man, Daniel Lambert, from his birth to the moment of his dissolution; with an account of men noted for their corpulency, and other interesting matter*, 26–27. New York: Samuel Wood & Sons. https://wellcomecollection.org/works/xycvjyp8.

Coe, T. 1752. A Letter from Dr. T Coe, Physician at Chelmsford in Essex, to Dr. Cromwell Mortimer, Secr. R. S. Concerning Mr Bright, the Fat Man at Malden in Essex. *Philosophical Transactions* 46:188–193. https://doi.org/10.1098/rstl.1751.0028.

M. Wübbels, *Quantification and Measurement in British Health Advice, c. 1670–1830*, Palgrave Studies in Medieval and Early Modern Medicine,
https://doi.org/10.1007/978-3-032-32760-4

Cullen, William. Case Note Regarding Anonymous, April 26 (Doc. ID:844), 3. 1773. The Consultation letters of William Cullen. Royal College of Physicians at Edinburg. accessible online via *The Cullen Project*. http://www.cullenproject.ac.uk.

Cullen, William, to Webster, Jane, September 16 (Doc. ID: 1021). 1780. The Consultation Letters of William Cullen. Royal College of Physicians at Edinburg. accessible online via *The Cullen Project*. http://www.cullenproject.ac.uk.

Daily Advertiser (London, England), Friday, May 6, 1796a, Issue 21038. 17th–18th Century Burney Collection Newspapers.

Daily Advertiser. 1796b. Issue 21038, Friday, May 6, London. 17th–18th Century Burney Collection Newspapers.

Merlin, John Joseph. 1787. Pamphlet featuring the 'Sanctorius Balance'. *Morning and Evening Amusements at Merlin's Mechanical Museum, No. 11 Princess Street, Hannover Square*. London. https://archive.org/details/b30375642/page/6.

Morning Herald. 1788. London, England, Friday, April 25, Issue 2343, 17th–18th Century Burney Collection Newspapers.

Morning Post. 1793a. London, England, Tuesday, December 17, Issue 6445. 17th–18th Century Burney Collection Newspapers.

Morning Post. 1793b. Issue 6445, Tuesday, December 17. London. 17th–18th Century Burney Collection Newspapers.

Morning Post and Fashionable World. 1796a. London, England, Saturday, May 7; Issue 7546. 17th–18th Century Burney Collection Newspapers.

Morning Post and Fashionable World. 1796b. Issue 7546, Saturday, May 7. London, 17th–18th Century Burney Collection Newspapers.

Webster, Jane, to Cullen, William, September 11. (Doc. ID: 1914). 1780. The Consultation Letters of Dr William Cullen (1710–1790). Royal College of Physicians at Edinburg. accessible online via *The Cullen Project*. http://www.cullenproject.ac.uk.

The Case of Mr. Thomas Wood, a Miller, of Billericay, in the County of Essex. 1771. *Medical Transactions*, 2, 259–274. Google Books.

A Sequel to the Case of Mr. Thomas Wood, of Billericay, in the County of Essex. 1785. *Medical Transactions*, 3, 309–318. Google Books.

II: Archival and Manuscript Sources

A collection of 8 'weighing' ledgers is currently kept at 6 Berry Bros. and Rudd, a wine merchant on No. 3 St. James Street, London.

Bayly, John, Case Book of John Bayly, *27 Sept. 1762 – 1 Sept. 1764*. MS105. Royal College of Physicians Archives.

A Collection of Letters written to and from Thomas Wood, 1769–1782. (Ref: D/DU1188). Essex Record Office.

III: Printed Sources

A.B. 1674. *The sick-mans rare jewel wherein is discovered a speedy way how every man may recover lost health, and prolong life*. London: Printed by T.R. and N.T. Early English Books Online.

Adair, J. M. 1804. *An Essay on Diet and Regimen as indispensable to the recovery and enjoyment of firm health*. London: Printed for James Ridgeway. Archive.org.

Anon. 1716. *The British Physician: Treating of out diet, and Common Nourishment, Of Air, Of Medicines, of Physical and Chyrurgical Writers, &c*. London: Printed for K. Morphew. Eighteenth Century Collections Online.

Anon. 1752. *The Best and easiest method of preserving interrupted health to extreme old age*. London: Sold by R. Baldwin. Eighteenth Century Collections Online.

Anon. 1772. *Directions and observations relative to food, exercise, and sleep*. London: Printed for S. Bladon. Eighteenth Century Collections Online.

Anon. 1827. *Sure Methods of improving health and prolonging life …and maxims of health, for the bilious and nervous, the consumptive, men of letters, and people of fashion, by a physician*. 2nd ed. London: Published by Simkin and Marshall, Stationer's Court. www.archive.org.

Arbuthnot, John. 1731. *An essay concerning the nature of aliments, and the choice of them, according to the different constitutions of human bodies*. Dublin: Printed for S. Powell. Eighteenth Century Collections Online.

Arbuthnot, John. 1732. *Practical rules of diet in the various constitutions and diseases of human bodies*. London: Printed for J. Tonson. Eighteenth Century Collections Online.

Archer, John. 1671. *Every man his own doctor in two parts, shewing how every man may know his own constitution by certain signs, also the nature and faculties of all food as well as meats as drinks*. London: Printed by Peter Lillicrap for the author. Early English Books Online.

Barry, Edward. 1759. *A treatise on the three different digestions, and discharges of the human body*. London: Printed for A, Millar. Eighteenth Century Collections Online.

Bennet, Timothy. 1734. *An essay on the gout; in which a method is propos'd to relieve the hereditary, and to cure the acquired … Such as have been reliev'd, or actually cur'd of the Gout by it*. London, printed for Richard Ford. Eighteenth Century Collections Online.

Brooke, Humphrey. 1650. *Hygiene, or a conservatory of health*. London: Printed by R.W. for G. Whittington. Early English Books Online.

Browne, Richard. 1729. *Medicina musica: Or, a mechanical essay on the effects of singing, musick, and dancing, on human bodies.* London: Printed for John Cooke. Eighteenth Century Collections Online.

Buchan, William. 1772. *Domestic medicine; or, a treatise on the prevention and cure of diseases by regimen and simple medicines*, 2nd ed. London: Printed for W. Strahan. Eighteenth Century Collections Online.

Burton, Robert. 1652. *The anatomy of melancholy what it is, with all the kinds, symptomes, prognostickes & severall cures of it*, 2nd ed. London: Printed by Hen. Crips & Lodo. Early English Books Online.

Burton, John. 1738. *A treatise on the non-naturals, in which the great influence they have on human bodies is set forth, and mechanically accounted for.* London: Printed by A. Staples. Eighteenth Century Collections Online.

Byfield, T. 1685. *Two discourses one of consumptions, with their cure by a new method, the other contains some rules of health.* London: Printed by Dorman Newman. Early English Books Online.

Byfield, T. 1717. *Directions tending to health and long life &c.* London: Printed for the Author. Eighteenth Century Collections Online.

Charleton, Walter. 1659. *Natural history of nutrition, life and voluntary motion containing all the new discoveries of anatomist's and most probably opinions of physicians.* London: Printed for Henry Herringman. Early English Books Online.

Cheyne, George. 1720. *Observations concerning the Nature and due Method of treating the Gout.* London, printed for G. Strahan. Eighteenth Century Collections Online.

Cheyne, George. 1724. *An Essay of health and long life.* London: Printed for George Strahan. Eighteenth Century Collections Online.

Cheyne, George. 1733. *The English malady: Or, a treatise of nervous diseases of all kinds.* London: Printed for G. Strahan. Eighteenth Century Collections Online.

Cheyne, George. 1740. *An essay on regimen.* London: Printed for C. Rivington. Eighteenth Century Collections Online.

Cheyne, George. 1742. *The natural method of cureing the diseases of the body, and the disorders of the mind depending on the body.* London: Printed for George Strahan. Eighteenth Century Collections Online.

Cocchi, Antonio. 1745. *The Pythagorean diet, of vegetables only, conductive to the preservation of health, and the cure of diseases.* London: Printed for R. Dodsley. Eighteenth Century Collections Online.

Cogan, Thomas. 1588. *The haven of health.* London: Printed by Melch Bradwood.

Cole, William, 1812. 1759, from John Nichols, *Literary Anecdotes of the Eighteenth Century*, 1. London: Nichols, Son, and Bentley.

Cornaro, Luigi. 1702. *Sure and Certain methods of attaining a long and healthful life, with means of correcting a bad constitution, &c. Written originally in Italian by Lewis Cornaro, a Noble Venetian … by W. Jones, A.B.* London: Printed for Thomas Leigh. Eighteenth Century Collections Online.

Cornaro, Luigi. 1740. *Sure and certain methods of attaining a long and healthful life, with means of correcting a bad constitution, &c. Written originally in Italian, by Lewis Cornaro...To which is prefix'd, Mr Addison's account and recommendation of this book.* Dublin: Printed for Richard Gunne. Eighteenth Century Collections Online.

Cornwell, Bryan. 1784. *The domestic physician; or, guardian of health.* London: J. Murray, J. Bew, L. Davis. Eighteenth Century Collections Online.

Cullen, William. 1784. *First Lines of the practice of physic*, Vol. 4. Edinburgh: Printed for C. Elliott. Archive.org.

Culpeper, Nicholas. 1659. *Culpeper's School of physick, or, The experimental practice of the whole art wherein are contained all inward diseases from the head to the foot.* London: Printed for N. Brook. Early English Books Online.

Curteis, Thomas. 1704. *Essays on the preservation & recovery of health.* London: Printed by Richard Wilkin. Eighteenth Century Collections Online.

De Valagin, Francis. 1768. *A treatise on diet, or, the management of human life; by physicians called the six non-naturals.* London: Printed for the author by J. and W. Oliver. Eighteenth Century Collections Online.

Democritus Junior. "Exercise rectified of Body and Mind", *The Anatomy of Melancholy.* London: Printed & to be sold by Hen. Crips & Lodo Lloyd. Project Gutenberg Online Version, The British Library Online: http://www.gutenberg.org/files/10800/10800-h/10800-h.htm.

Durante, Castor. 1686. *A Treasure of health by Castor Durante Da Gualdo, physician and citizen of Rome, wherein is shewn how to preserve health and prolong life.* London: Printed for William Crook. Early English Books Online.

Falconer, William. 1778. *Observations on some of the articles of diet and regimen usually recommended to valetudinarians.* London: Printed for Edward and Charles Dilly. Eighteenth Century Collections Online.

Falconer, William. 1781. *Remarks on the influence of climate, situation, nature of country, population, nature of food, and way of life, on the disposition and temper ... of mankind.* London, Printed for C. Dilly. Eighteenth Century Collections Online.

Flamant, M.. 1697. *The art of preserving and restoring health explaining the nature and causes of the distempers that afflict mankind, also shewing that every man is, or may be his own physician.* London: Printed by R. Bently, H. Bonwich, and S. Manship. Early English Books Online.

Flemyng, Malcolm. 1760. *A discourse on the nature, causes and cure of corpulency.* London: Printed for L. Davis and C. Reymers. Eighteenth Century Collections Online.

Floyer, John. 1725. *Medicina gerocomica: Or, the galenic art of preserving old men's health explain'd.* London: Printed for J. Isted. Eighteenth Century Collections Online.

Forster, William. 1738. *A treatise on the various kinds and qualities of foods: With aphorisms of health, or, rules to preserve the body to a good age*. Newcastle upon Tyne: Printed by John White. Eighteenth Century Collections Online.

Fothergill, John. 1762. *Rules for the preservation of health: Containing all that has been recommended by the most eminent physicians*. London: Printed for John Pridden. Eighteenth Century Collections Online.

Fothergill, John. 1781. Case of an Angina Pectoris. In *A Complete Collection of the Medical and Philosophical works of John Fothergill*, edited by John Elliot. London: Printed for John Walker. https://wellcomecollection.org/works/ugtus2gc.

Freind, John. 1752. *Emmenologia: Written, in Latin, by the late learned Dr. Freind, translated into English by Thomas Dale*, 2nd ed. London: Printed for T. Cox. Eighteenth Century Collections Online.

Fuller, Francis. 1705. *Medicina Gymnastica: Or, a treatise concerning the power of exercise*. London: Printed by John Matthews. Eighteenth Century Collections Online.

Graham, T. J. 1827. *Sure Methods of Improving Health and Prolonging Life; Or, A Treatise on the Art of Living Long and Comfortably by Regulating The Diet and Regimen*. London: Published for the author by Simpkin and Marshall. Early English Books Online.

Grosvenor, Benjamin. 1716. *Health, An essay on its nature, value, uncertainty, and best improvement*. London: Printed for E. Matthews. Eighteenth Century Collections Online.

Hartman, G. 1682. *The true preserver and restorer of health being a choice of select and experienced remedies for all distempers incident to men, women, and children*. London: Printed by T.B. for the author. Early English Books Online.

Harvey, Gideon. 1702. *The third edition of the Vanities of philosophy and physick … with variety of Medicines and Rules, whereby to make particular Choice out of them*, 3rd ed. London: Printed for A. Roper. Eighteenth Century Collections Online.

Harwood, Edward. 1774. *Of temperance and intemperance: Their effects on the body and mind, and their influence in prolonging or abbreviating human life*. London: Printed for T. Becket. Eighteenth Century Collections Online.

Hill, John. 1750. *The old man's guide to health and longer life: With rules for diet, exercise, and physick*, 2nd ed. London: Printed for M. Cooper. Eighteenth Century Collections Online.

Hill, John. 1758. *The management of the gout, by a physician from his own case*, 3rd ed. London: Printed for R. Baldwin. Eighteenth Century Collections Online.

Hufeland, Christopher. 1829. *The art of prolonging human life*. London: Published by Simkin and Marshall, Stationer's Court.

Hygiasticon; or, a treatise of the means of health and long life, written originally in Latin by Leonardus Lessius, Now rendered into English by T. Smith, Whereunto is annexed, Cornaro's treatise of the benefits of a Sober life. 1742. London: Printed for the Author and sold by Charles Hitch. Eighteenth Century Collections Online.

Jameson, Thomas. 1811. *Essays on the changes of the human body at its different ages.* London: Printed for Longman, Hurst, Rees, Orme, and Browne. Archive.org.

Keill, James. 1738. *Essays on the several parts of the animal oeconomy.* London: Printed for George Strahan. Eighteenth Century Collections Online.

Lessius, Leonardus. 1636. *Hygiasticon: Or, The right course of preserving life and health unto extream old age together with soundnesse and integritie of the senses, judgement, and memorie.* 2nd edition. Cambridge: Printed by R. Daniel and T. Buck. Early English Books Online.

Lynch, Bernard. 1744. *A guide to health through the various stages of life.* London: Printed for the author, and sold by Mrs. Cooper. Eighteenth Century Collections Online.

Mackenzie, James. 1758. *The history of health, and the Art of preserving it.* Edinburgh: Printed and sold by William Gordon. Eighteenth Century Collections Online.

Makluire, John. 1630. *The Buckler of bodilie health whereby health may bee defended and sicknesse repelled*, 2nd ed. Edinburgh: Printed by John Wreittoun. Early English Books Online.

Maynwaring, E. 1669. *Vita sana & longa the preservation of health and prolongation of life proposed and proved*, 55–56. London: Printed by J.D.

Maynwaring, Everard. 1683. *The method and means of enjoying health, vigour and long life adapting peculiar courses for different constitutions, ages, abilities, valetudinary states, individual proprieties, habituated customs, and passions of mind.* London: Printed by J.M. for Dorman Newman. Early English Books Online.

Medicus. 1830. *The oracle of health and long life, or plain rules for the attainment and preservation pf sound health and vigorous old age; with rational instructions for diet, Regimen &c.* London: Printed for Sherwood, Gilbert & Piper. www.archive.org.

Muffett, Thomas. 1655. *Healths improvement, or, rules comprizing and discovering the nature, method, and manner of preparing all sort of food used in this nation … corrected and enlarged by Christopher Bennet.* London: Printed by Thomas Newcomb. The British Library.

Pepys, Samuel, 2000. September 2, 1667. *The Diary of Samuel Pepys*, Vol. 8: 1667. Edited by Robert Latham and William Matthews. London: HarperCollins.

Priestley, Robert. 1798. *A few interesting remarks on bilious disorders, particularly the jaundice, and yellow fever, &c. Highly important to persons of fashion, the sedentary, and those who inhabit or frequent tropical climates; recommending the anti-bilious powder, a certain and experienced remedy.* London, Printed by Thomas Gill.

Ramazzini, Bernardino. 1705. *A treatise of the diseases of tradesmen, shewing the various influence of particular trades upon the state of health.* London: Printed for Andrew Bell et al. Eighteenth Century Collections Online.

Robinson, Bryan. 1747. *A dissertation on the food and discharges of human bodies.* Dublin: Printed by S. Powell. Eighteenth Century Collections Online.

Santori, Santorio. 1676. *Medicina Statica, or, Rules of health in eight sections of aphorisms originally written by Sanctorius; English'd by J.D.* London: Printed for John Starkey. Early English Books Online.

Santori, Santorio. 1712. *Medicina Statica: Being the aphorisms of Sanctorius, translated into English with large explanations. Wherein is given a mechanical account of the animal economy, and of the Efficacy of the Non-Naturals, ... by John Quincy.* London: Printed for William Newton. Eighteenth Century Collections Online.

Santori, Santorio. 1720. *Medicina Statica: Being the aphorisms of Sanctorius ... To which is added Dr. Keill's Medicina Statica Britannica, with comparative remarks ... by John Quincy.* 2nd edition. London: Printed for W. and J. Newton. Eighteenth Century Collections Online.

Santori, Santorio. 1737. *Medicina Statica: Being the aphorisms of Sanctorius, translated into English, with large explanations. To which is added, Dr. Keill's Medicina Statica Britannica ... By John Quincy.* 5th edition. London: Printed for T. Longman. Eighteenth Century Collections Online.

Scudamore, Charles. 1819. *A treatise on the nature and cure of gout and rheumatism.* London: Longman, Rees, Orme and Brown. Archive.org.

Sharp, Jane. 1671. *The midwives book, or, The whole art of midwifery discovered.* 2nd edition. London: Printed for Simon Miller. Early English Books Online.

Short, Thomas. 1727. *A discourse concerning the causes and effects of corpulency: Together with the methods for its prevention and cure.* London: Printed for J. Roberts. Eighteenth Century Collections Online.

Short, Thomas. 1750. *New observations of city, town and country bills of mortality.* London: Printed for T. Longman. Archive.org.

Smith, William. 1776. *A sure guide in sickness and health, in the choice of food, and use of medicine.* London: Printed for and sold by J. Bew. Eighteenth Century Collections Online.

Spilsbury, Francis. 1788. *Free observations on the scurvy, gout, diet, and remedy.* London: W. Gillman. Eighteenth Century Collections Online.

Stark, William. 1788. *The works of the late William Stark, M.D. consisting of clinical and anatomical observations, with experiments, dietetical and statical, revised and published from his original Mss by James Carmichael Smyth.* London: Printed for J. Johnson. Eighteenth Century Collections Online.

Stephens, William. 1732. *Dolaeus upon the cure of the gout by Milk-Diet, to which is prefixed, an essay upon Diet.* London: Printed for J. Smith and W. Bruce. www.archive.org.

Strother, Edward. 1719. *The family companion for health, or, plain, easy and certain rules.* London: Printed for F. Fayram. Eighteenth Century Collections Online.

Strother, Edward. 1725. *An essay on sickness and health; wherein are contain'd all necessary cautions and directions.* London: Printed for Charles Rivington. Eighteenth Century Collections Online.

Taylor, Joseph. 1818. *Annals of health and long life: With observations on regimen and diet, so necessary to the preservation of life.* London: Printed for Effingham Wilson. https://wellcomecollection.org/works/vexhyptv.

Temple, William. 1701. *Miscellanea, The third part, containing I: An essay on popular discontents. II: An essay upon health and long life.* London: Printed by Benjamin Tooke. Eighteenth Century Collections Online.

The Letters of Doctor George Cheyne to Samuel Richardson (1733–1743). 1943. Edited by Mullett, Charles, F., Columbia: University of Missouri.

The temperate man, or the right way of preserving life and health ...in three treatises, the first written by the learned Leonardus Lessius, The second by Lodowick Cornaro, a Noble Gentleman of Venice, the Third by a famous Italian. 1678. London, printed by J.R. for John Starkey. Early English Books Online.

The universal family physician, and surgeon. Containing a familiar and accurate description of the symptoms of every disorder incident to Mankind; together with their gradual Progress, and method of cure. 1798. London: Printed by Nemingway and Nuttall. Archive.org.

Tissot, Samuel Auguste. 1766. *Advise to the people in general, with regard to their health. Translated from the French edition of Dr. Tissot's Avis au people &c. Printed at Lyons, with all the notes in the former English editions, and a few additional ones, by J. Kirkpatrick.* DublinL: Printed for J. Hoey. Wellcome Collection Online.

Trusler, John. 1775. *An easy way to prolong life, by a little attention to what we eat and drink.* London: Printed for the author, and sold by J. Bell. Eighteenth Century Collections Online.

Tryon, Thomas. 1691. *The way to health, long life and happiness.* 2nd edition. London: Printed by H.C. for D. Newman. Early English Books Online.

Tryon, Thomas. 1692. *The good house-wife made a doctor, or, Health's choice and sure friend being a plain way of nature's own prescribing to prevent and cure most diseases incident to men, women and children by diet and kitchin-physick only.* London: Printed for H.N. and T.S.. Early English Books Online.

Vaughan, William. 1612. *Approved directions for health, both naturall and artificiall derived from the best physitians as well moderne as auncient.* 2nd edition. London: Printed by T. S[nodham] for Roger Lackson. Early English Books Online.

Wainewright, Jeremiah. 1707. *A mechanical account of the non-naturals: Being a brief explication of the changes made in humane bodies, by air, diet &c.* London: Printed for Ralph Smith. Eighteenth Century Collections Online.

Webster, Joshua. 1804. *Practical observations on the preservation of health and instructions for the preventions and cure of human maladies.* London: J. Slee. https://wellcomecollection.org/works/rsktaxjp.

Willich, Anthony. 1799. *Lectures on diet and regimen: Being a systematic inquiry into the most rational means of preserving health and prolonging life.* London: Printed for T.N. Longman and O. Rees. Eighteenth Century Collections Online.

Secondary Sources

Ajana, Btihaj, Joaquim Braga, and Simone Guidi. 2021. Introduction. In *The quantification of bodies in health: Multidisciplinary perspectives*, edited by Btihaj Ajana, Joaquim Braga, and Simone Guidi, 1–12. Leeds: Emerald Publishing Limited. https://doi-org.uoelibrary.idm.oclc.org/10.1108/978-1-80071-883-820211001.

Ajana, Btihaj. 2018. Introduction: Metric culture and the over-examined life. In *Metric culture*, edited by Btihaj Ajana, 1–9. Leeds: Emerald Publishing Limited. https://doi-org.uoelibrary.idm.oclc.org/10.1108/978-1-78743-289-520181001.

Albala, Ken. 2002a. *Eating right in the renaissance.* London: University of California Press.

Albala, Ken. 2003. *Food in early modern Europe.* Westport: Greenwood Publishing Group.

Albala, Ken. 2012. Food for healing: Convalescent cookery in the early modern era. *Studies in History and Philosophy of Biological and Biomedical Sciences* 42 (2): 323–328.

Albala, Ken. 2002b. Insensible perspiration and oily vegetable humor: An eighteenth century controversy over vegetarianism. *Gastronomica* 2 (3): 29–36.

Albala, Ken. 2005. Weight loss in the age of reason. In *Cultures of the abdomen: Diet, digestion, and fat in the modern world*, edited by Christopher E. Forth and Ana Carden-Coyne, 169–184. Basingstoke: Palgrave Macmillan.

Antonelli, Francesca. 2022. Weighing authority: Lavoisier's and Séguin's reassessment of Santorio's experiments on transpiration. In *Santorio Santori and the emergence of Quantified Medicine, 1614–1790*, edited by Jonathan Barry and Fabrizio Bigotti, 371–391. London: Palgrave Macmillan.

Arcangeli, Alessandro. 2016. Exercise for women. In *Sports and physical exercise in early modern culture*, edited by Rebekka Von Mallinckrodt and Angela Schattner, 147–164. Abingdon: Routledge.

Arcangeli, Alessandro. 2003. *Recreation in the renaissance: Attitudes towards leisure and pastimes in European Culture, c. 1425–1675*. Basingstoke: Palgrave Macmillan.

Barry, Jonathan. 2018. The “Compleat Physician” and experimentation in medicines: Everard Maynwaring (c.1629–1713) and the restoration debate on medical practice in London. *Medical History* 62 (2): 155–176. https://doi.org/10.1017/mdh.2018.2.

Behringer, Wolfgang. July 2009. Arena and pall mall: Sport in the early modern period. *German History* 27 (3): 331–357. https://doi.org/10.1093/gerhis/ghp030.

Berryman, Jack. January 1989. The tradition of the six things non-natural: Exercise and medicine from Hippocrates through Ante-Bellum America. *Exercise and Sport Sciences Reviews* 17 (1): 515–560.

Bigotti, Fabrizio. 2023. Intensity meters: New notes and discoveries on the invention of early modern precision instruments. *Early Science and Medicine* 28:597–630. https://doi.org/10.1163/15733823-20230088.

Bigotti, Fabrizio, and Jonathan Barry. 2022. Introduction. In *Santorio Santori and the emergence of quantified medicine, 1614–1790*, edited by Jonathan Barry and Fabrizio Bigotti, 1–63. London: Palgrave Macmillan.

Bigotti, Fabrizio. 2018. The weight of the Air: Santorio's thermometers and the early history of medical quantification reconsidered. *Early Modern Studies* 7 (1): 73–103.

Blackman, Cally. 2001. Walking amazons: The development of the riding habit in England during the Eighteenth Century. *Costume* 35 (1): 47–58. https://doi.org/10.1179/cos.2001.35.1.47.

Bohstedt, John. 2010. *The politics of provisions: Food riots, moral economy, and market transition in England, c.1550–1850*. Ashgate: Farnham.

Bondeson, Jan. 2006. *Freaks: The pig-faced lady of Manchester Square and other medical marvels*. Stroud: Tempus Publishing.

Booth, Alan. November 1977. Food riots in the North-West of England, 1790–1801. *Past and Present* 77:84–107.

Bourg, Jean-Francois, and Jean-Jacques Gouguet. 2005. *Economie du Sport*. Paris: Presses Univertaires de France.

Brailsforth, Dennis. Winter 1982. Sporting days in eighteenth century England. *Journal of Sport History* 9 (3): 41–54.

Brown, Theodore M. Autumn 1974. From mechanism to vitalism in Eighteenth-Century English physiology. *Journal of the History of Biology* 7 (2): 179–216.

Bylebyl, Jerome J. 1977. Nutrition, quantification and circulation. *Bulletin of the History of Medicine* 51 (3): 369–385.

Carroll, P. E. 2002. Medical police and the history of public health. *Medical History* 46 (4): 461–494.

Cavallo, Sandra, and Tessa Storey. 2013. *Healthy living in late renaissance Italy*. Oxford: Oxford University Press.

Cavallo, Sandra, and Tessa Storey. 2016. 'Decorous' and pleasant exercise: Competing models and the practices of the Italian Nobility. In *Sports and physical exercise in early modern culture*, edited by Rebekka Von Mallinckrodt and Schattner Angela, 165–188. Abingdon: Routledge.

Cavallo, Sandra. 2017. Introducing Conserving health: The Non-Naturals in early modern culture and society. In *Conserving health in early modern Europe: Bodies and environments in Italy and England*, edited by Sandra Cavallo and Tessa Storey, 1–20. Manchester: Manchester University Press.

Churchill, Wendy D. 2005. The medical practice of the sexed body: Women, men, and disease in Britain, circa 1600–1740. *Social History of Medicine* 18 (1): 3–22.

Clericuzio, Antonio. 2012. Chemical and mechanical theories of digestion in early modern medicine. *Studies in History and Philosophy of Biological and Biomedical Sciences* 43:329–337.

Coleman, William. October 1974. Health and hygiene in the 'Encyclopedie': A medical doctrine for the bourgeoisie. *Journal of the History of Medicine and Allied Sciences* 29 (4): 399–421.

Collins, Tony. 2007. Review article: Work rest and play: Recent trends in the history of sport and leisure. *Journal of Contemporary History* 42 (2): 397–410.

Dacome, Lucia. 2005. 'Useless and pernicious matter': Corpulence in eighteenth-century England. In *Cultures of the Abdomen: Diet, digestion, and fat in the modern world*, edited by Christopher E. Forth and Ana Carden-Coyne, 185–204. Basingstoke: Palgrave Macmillan.

Dacome, Lucia. 2012. Balancing act. *Studies in History and Philosophy of Biological and Biomedical Sciences* 43:379–391.

Dacome, Lucia. 2001. Living with the chair: Private excreta, collective health and medical authority in the eighteenth century. *History of Science* 39:467–500.

De Renzi, Sylvia. 2004. Women and Medicine. In *The healing arts: Health, disease and society in Europe, 1500–1800*, edited by Peter Elmer, 166–195. Manchester: Manchester University Press.

Dinges, Martin. 2016. Exercise, health and gender: Normative discourses and practices in eighteenth- and nineteenth-century German-Speaking countries. In *Sports and physical exercise in early modern culture*, edited by Rebekka Von Mallinckrodt and Schattner Angela, 189–208. Abingdon: Routledge.

Donato, Maria Pia. 2017. The afterlife of the non-naturals in early eighteenth-century Hippocratism: From the healthy individual to the healthy population. In *Conserving health in early modern Europe: Bodies and environments in Italy and England*, edited by Sandra Cavallo and Tessa Storey, 158–182. Manchester: Manchester University Press.

Duden, Barbara. 1991. *The Woman Beneath the skin: A doctor's patients in eighteenth-century Germany*. Translated by Thomas Dunlap. Cambridge: Harvard University Press.

Earle, Rebecca. February 2019. The political economy of nutrition in the eighteenth century. *Past and Present* 242 (1): 79–117.

Edwards, Peter. 2007. *Horse and man in early modern England*. London: Bloomsbury. http://dx.doi.org.uoelibrary.idm.oclc.org/10.5040/9781474210072.ch-005.

Emch-Deriaz, Antoinette. 1992. The non-naturals made easy. In *The popularization of medicine, 1650–1850*, edited by Roy Porter, 134–159. Abingdon: Routledge.

Forth, Christopher E. 2019. *Fat: A cultural history of the stuff of life*. London: Reaktion Books.

Frangsmyr, Tore, J. L. Heilbron, and Robin E. Rider. 1990. *The quantifying spirit in the eighteenth century*. Berkely: University of California Press.

Gentilcore, David. 2016. *Food and health in early modern Europe: Diet, medicine and society, 1450–1800*. London: Bloomsbury.

Gowrley, Freya. December 2021. Size matters. *History Today* 71 (12): 42–47.

Guerrini, Anita. 2012. Health, national character and the English diet in 1700. *Studies in History and Philosophy of Biological and Biomedical Sciences* 43:349–356.

Guerrini, Anita. 2020. The impossible ideal of moderation: Food, drink, and longevity. In *Lifestyle and medicine in the enlightenment*, edited by James Kennaway and Rina Knoeff, 86–108. London: Routledge. https://doi.org/10.4324/9780429465642.

Guerrini, Anita. 2000. *Obesity and depression in the enlightenment: The life and times of George Cheyne*. Norman: University of Oklahoma Press.

Guidi, Simone, and Joaquim Braga. 2023. Introduction. In *The quantification of life and health from the sixteenth to the Nineteenth century: Intersections of medicine and philosophy*, Palgrave Studies in Medieval and Early Modern Medicine, Palgrave Studies in Medieval and Early Modern Medicine, edited by Simone Guidi and Joaquim Braga. London: Palgrave Macmillan Cham. https://doi.org/10.1007/978-3-031-15725-7.

Guttmann, Allen. 1978. *From ritual to record: The nature of modern sports*. New York: Columbia University Press.

Handley, Sasha. 2016. *Sleep in early modern England*. New Haven: Yale University Press.

Harris, Bernard, Roderick Floud, and Sok Chul Hong. 2015. How many calories? food availability in England and Wales in the eighteenth and nineteenth centuries. *Research in Economic History* 31:111–191.

Hill, Steward. 2016. *Leisure cultures in urban Europe, c. 1700–1870*. Manchester: Manchester University Press.

Houlihan Flynn, Carol. 1990. Running out of matter: The body exercised in eighteenth-century fiction. In *The languages of psyche*, edited by G. S. Rousseau, 147–185. Oakland: University of California Press.

Huggins, Mike. 2017. Early modern sport. In *The Oxford handbook of sports history*, edited by Robert Edelman and Wayne Wilson, 113–128. Oxford: Oxford University Press.

Ishizuka, Hisao. 2012. 'Fibre body': The concept of fibre in eighteenth-century medicine, c.1700–40. *Medical history* 56 (4): 562–584. https://doi.org/10.1017/mdh.2012.74.

Ishizuka, Hisao. 2016. *Fiber, medicine, and culture in the British Enlightenment*. New York: Palgrave Macmillan.

Jarcho, Saul. July–August 1970. Galen's six non-naturals: A bibliographic note and translation. *Bulletin of the History of Medicine* 44, 4:372–377.

Johnes, Martin. 2008. British sports history: The present and the future. *Journal of Sports History* 35 (1): 65–71.

Kaplan, Steven L. 1976. *Bread, politics and political economy in the Reign of Louis XV*. Vol. 2. Dordrecht: Springer.

Kennaway, James, and Rina Knoeff. 2020. 'For it is the debilitating fibres that exercise restores': Movement, morality and moderation in eighteenth-century medical advice literature. In *Lifestyle and medicine in the enlightenment*, edited by James Kennaway and Rina Knoeff, 111–138. London: Routledge. https://doi.org/10.4324/9780429465642.

Kuriyama, Shigehisa. Fall 2008. The forgotten fear of excrement. *Journal of Medieval and Early Modern Studies* 38 (3): 413–442. https://doi.org/10.1215/10829636-2008-002.

Lane, Joan. 1985. 'The doctor scolds me': The diaries and correspondence of patients in eighteenth-century England. In *Patients and practitioners: Lay perceptions of medicine in pre-industrial society*, edited by Roy Porter, 205–248. Cambridge: Cambridge University Press.

Laqueur, Thomas. 1992. *Making sex: Body and gender from the Greeks to Freud*. London: Harvard University Press.

Levy-Navarro, Elena. 2010. Changing conceptions of the fat body in western history. In *Historicizing fat in Anglo-American Culture*, edited by Elena Levy-Navarro, 1–18. Columbus: Ohio State University Press.

Lious-Courvoisier, Micheline, and Severine Pilloud. 2004. Consulting by letter in the eighteenth century: Mediating the Patients's View? In *Cultural approaches to the history of medicine: Mediating medicine in early modern and modern Europe*, edited by Willem de Blecourt and Comelie Usborne, 71–88. Basingstoke: Palgrave Macmillan.

Madella, Laura. 2023. The more the year the less the food: Alvise Cornaro on *The Sober Life* (1558). In *The Quantification of life and health from the sixteenth to the Nineteenth century: Intersections of Medicine and Philosophy*, Palgrave Studies in Medieval and Early Modern Medicine, Palgrave Studies in Medieval and Early Modern Medicine, edited by Simone Guidi and Joaquim Braga. London: Palgrave Macmillan Cham. https://doi.org/10.1007/978-3-031-15725-7.

McClelland, John. 2009. *and Merrilees, Brian, Sport and culture in early modern Europe*. Toronto: Centre for Reformation and Renaissance Studies, Victory University of Toronto.

McClelland, John. 2007. *Body and Mind: Sport in Europe from the Roman Empire to the Renaissance*. Abingdon: Routledge.

Meredith, David, and Deborah Oxley. February 2014. Food and fodder: Feeding England, 1700–1900. *Past and Present* 222:163–214.

Mikkeli, H. 1999. *Hygiene in the early modern medical tradition*. Helsinki: Finnish Academy of Science and Letters.

Muldrew, Craig. 2011. *Food, energy and the creation of industriousness: Work and material culture in Agrarian England, 1550–1780*. Cambridge: Cambridge University Press.

Niebyl, Peter H. 1971. The non-naturals. *Bulletin for the History of Medicine* 45 (5): 486–492.

O'Hara-May, Jane. Autumn 1971. Measuring man's needs. *Journal of the History of Biology* 4 (2): 249–273.

Olsen, Kirsten. 2017. *Daily life in 18th-century England*. Westport: Greenwood Press.

Orland, Barbara. 2012. The fluid mechanics of nutrition: Herman Boerhaave's synthesis of seventeenth-century circulation physiology. *Studies in History and Philosophy of Biological and Biomedical Sciences* 43:357–369.

Orland, Barbara, and E. C. Spary. 2012. Introduction. *Studies in History and Philosophy of Biological and Biomedical Sciences* 43:317–322.

Packham, Catherine. 2012. *Eighteenth-century vitalism: Bodies, culture, politics*. Basingstoke: Palgrave Macmillan.

Pender, Stephen. 2014. Heat and moisture, rhetoric and spiritus. *Intellectual History Review* 24 (1): 89–112.

Pomata, Gianna. 2018. Innate heat, radical moisture and generation. In *Reproduction: Antiquity to the present day*, edited by Nick Hopwood, Rebecca Fleming, and Lauren Kassell, 195–208. Cambridge: Cambridge University Press.

Pomata, Gianna. 2001. Menstruating men: Similarity and difference of the sexes in early modern medicine. In *Generation and degeneration: Tropes of reproduction in literature and history from antiquity through early modern Europe*, edited by Valeria Finucci and Kevin Brownlee. Durham, NC: Duke University Press. https://doi.org/10.1215/9780822380276-005.

Porter, Roy. 2003. *Flesh in the age of reason*. London: Penguin Books.

Porter, Roy. 1985. *Patients and practitioners: Lay perceptions of medicine in pre-industrial society*. Cambridge: Cambridge University Press.

Purnis, Jan. 2023. Sanctorius's weighing chair: Measurement, metabolism, and mind. In *The Quantification of life and health from the sixteenth to the Nineteenth century: Intersections of Medicine and Philosophy*, Palgrave Studies in Medieval and Early Modern Medicine, Palgrave Studies in Medieval and Early Modern Medicine, edited by Simone Guidi and Joaquim Braga. London: Palgrave Macmillan Cham. https://doi.org/10.1007/978-3-031-15725-7.

Pyrges, Alexander. Forthcoming. 'No age did ever afford more instances of corpulency'. Obesity as a collective condition and the early modern medicalization of Girth. In *Fat Bodies in Early Modern Europe.* Edited by J. Burke, H. Fletcher, C. Ott.

Rankin, Alisha. 2013. *Panaceia's daughters: Noblewomen as healers in early modern Germany.* Chicago: Chicago University Press.

Rather, L. J. 1968. The 'six things non-natural': A note on the origins and fate of a doctrine and a phrase. *Clio Medica* 3:337–347.

Reinke-Williams, Tim. 2014. *Women, work and sociability in early modern London.* London: Palgrave Macmillan. https://doi.org/10.1057/9781137372109.

Renbourn, E. T. 1960. The natural history of insensible perspiration: A forgotten doctrine of health and disease. *Medical History* 4 (2): 135–152. https://doi.org/10.1017/s0025727300025229.

Rey, Roselyne. 1995. Vitalism, disease and society. In *Medicine in the enlightenment*, edited by Roy Porter, 274–287. Amsterdam: Rodopi.

Rogers, Pat. 2010. Fat is a fictional issue: The novel and the rise of weight-watching. In *Historicizing fat in Anglo-American culture*, edited by Elena Levy-Navarro, 19–42. Columbus: Ohio State University Press.

Rusnock, Andrea A. February 1995. The weight of evidence and the burden of authority: Case histories, medical statistics and smallpox inoculation. *Medicine in the Enlightenment* 29:289–315.

Rusnock, Andrea A. 2002. *Vital accounts: Quantifying health and population in eighteenth-century England and France.* Cambridge: Cambridge University Press.

Sachiko, Kusakawa. 2004. Medicine in Western Europe in 1500. In *The healing arts, health, disease and society in Europe, 1500–1800*, edited by Peter Elmer, 1–24. Manchester: Manchester University Press.

Schattner, Angela. June 2014. For the recreation of gentlemen and other fit persons of the better sort: Tennis courts and bowling greens as early leisure venues in sixteenth- to eighteenth-century London and Bath. *Sport in History* 34 (2): 198–222.

Schwartz, Hillel. 1986. *Never satisfied: A cultural history of diets, fantasies and fat.* New York: Anchor.

Shammas, Carole. July 1984. The eighteenth- century english diet and economic change. *Explorations in Economic History* 21 (3): 254–269.

Shapin, Steven. August 2014. 'You are what you eat': Historical changes in ideas about food and identity. *Historical Research* 87 (237): 377–392.

Shapin, Steven. 2003. Trusting George Cheyne: Scientific expertise, common sense and moral authority in early eighteenth-century dietetic medicine. *Bulletin of the History of Medicine* 77:263–297.

Shapin, Steven. March 5, 2018. Was Luigi Cornaro a dietary expert? *Journal of the History of Medicine and Allied Sciences* 73 (2): 135–149.

Sherman, Sandra. 2001. *Imagining poverty: Quantification and the decline of paternalism*. Columbus: Ohio State University Press.

Siena, Kevin. 2010. Pliable bodies: The moral biology of heath and disease. In *A cultural history of the human body: In the enlightenment*, edited by Carole Reeves, 33–52. Oxford: Berg.

Skuse, Alanna. 2015. *Constructions of cancer in early modern England*, Palgrave Studies in Literature, Science, and Medicine, 98–99. Basingstoke: Palgrave Macmillan.

Slack, Paul. 1979. *Mirrors of health and treasures of poor men: The use of vernacular medical literature in Tudor England*. Edited by Webster, , 237–273. Charles. Cambridge: Cambridge University Press.

Smith, Ginnie. 1985. Prescribing the rules of health: Self-help and advice in the late eighteenth century. In *Patients and practitioners: Lay perceptions of medicine in pre-industrial society*, edited by Roy Porter, 249–282. Cambridge: Cambridge University Press.

Spary, E. C. 2021. Food. In *A cultural history of medicine: In the enlightenment*, edited by Lisa Wynne Smith, 29–50. London: Bloomsbury.

Spary, E. C. 2012. *Eating the enlightenment: Food and the sciences in Paris*. Chicago: University of Chicago Press.

Stolberg, Michael. 2012a. 'Abhorreas pinguedinem': Fat and obesity in early modern medicine (c. 1500–1750). *Studies in History and Philosophy of Biological and Biomedical Sciences* 43:370–378.

Stolberg, Michael. June 2003. A woman down to her bones: The anatomy of sexual difference in the sixteenth and seventeenth centuries. *Isis* 94 (2): 274–299.

Stolberg, Michael. 2020. Keeping the body open: Impurity, excretions and healthy living in the early modern period. In *Lifestyle and medicine in the enlightenment*, edited by James Kennaway and Rina Knoeff, 205–222. London: Routledge. https://doi.org/10.4324/9780429465642.

Stolberg, Michael. 2012b. Sweat. Learned concepts and popular perceptions, 1500–1800. In *Blood, sweat and tears: The changing concepts of physiology from antiquiry into early modern Europe*. Edited by Horstmanshoff, Manfeld, King, Helen, and Zittel, Claus, 503–522. https://doi.org/10.1163/9789004229204_022.

Stolberg, Michael. 2011. *Experiencing illness and the sick body in early modern Europe*. London: Palgrave Macmillan.

Storey, Tessa. 2020. 'That is more excellent which preserveth health and preventeth sicknesse': Continuity and change in vernacular preventative health advice over the early modern period. In *Lifestyle and medicine in the enlightenment*, edited by James Kennaway and Rina Knoeff, 289–312. London: Routledge. https://doi.org/10.4324/9780429465642.

Sysling, Fenneke. July 2020. Measurement, self-tracking and the history of science: An introduction. *History of Science* 58 (2): 103–116. https://doi.org/10.1177/0073275319865830.

Szymanski, Stefan. Spring 2008. A theory of the evolution of modern sport. *Journal of Sports History* 35 (1): 1–32.

Temkin, Owsei. 1961. A Galenic model for quantitative physiological reasoning? *Bulletin of the History of Medicine* 35 (5): 470–475.

Temkin, Owsei. 1960. Nutrition from classical antiquity to the baroque. In *Human Nutrition, Historic and Scientific*, edited by Iago Galdston, 78–97. New York: International Universities Press.

Thomas, Raymond. 1991. *Histoire du Sport*. Paris: PUF.

Tomlinson, Alan, and Christopher Young. 2011. Towards a new history of European sport. *European Review* 19:487–507.

Toulalan, Sarah. February 2014. 'To[o] much eating stifles the child': Fat bodies and reproduction in early modern England. *Historical Research* 87 (235): 65–93. https://doi.org/10.1002/1468-2281.12031.

Tranter, Neil. 1998. *Sport, economy and society in Britain, 1750–1914*. Cambridge: Cambridge University Press.

Turner, Bryan. 1982. The government of the body: Medical regiments and the rationalisation of diet. *British Journal of Sociology* 33:254–269.

Turner, David. 2010. The body beautiful. In *A cultural history of the human body in the enlightenment*, edited by Carole Reeves. Oxford: Berg.

Vamplew, Wray. 2017. Industrialisation and sport. In *The Oxford handbook of sports history*, edited by Robert Edelman and Wayne Wilson, 131–144. Oxford: Oxford University Press. https://doi.org/10.1093/oxfordhb/9780199858910.013.17.

Verwaal, Ruben E. 2020a. *Bodily fluids, chemistry and medicine in the eighteenth-century boerhaave school*. EBook,. Palgrave Macmillan. https://doi.org/10.1007/978-3-030-51541-6.

Verwaal, Ruben. 2020b. Increasing and reducing: Breast milk flows and female health. In *Lifestyle and medicine in the enlightenment*, edited by James Kennaway and Rina Knoeff, 223–240. London: Routledge. https://doi.org/10.4324/9780429465642.

Vigarello, Georges. 2013. *The metamorphoses of fat: A history of obesity*. Columbia: Columbia University Press.

Von Mallinckrodt, Rebekka. 2004. Bewegte Geschichte – Pladoyer for eine verstarkte Integration und konzeptuelle Erweiterung der Sportgeschichte in der

fruhneuzeilichen Geschichtswissenschaft. *Historische Anthropologie* 12 (1): 134–139.
Von Mallinckrodt, Rebekka, and Angela Schattner. 2016. *Sport and physical exercise in early modern culture: New perspectives on the history of sport and motion.* Abingdon: Routledge.
Von Mallinckrodt, Rebekka. 2008. *Bewegtes leben. Korpertechniken in der fruhen neuzeit.* Wiesbaden: Harrassowitz.
Ward, Paul. January 2013. Last man picked. Do mainstream historians needs to play with sports historians? *International Journal of the History of Sports* 30 (1): 6–13.
Wear, Andrew. 1992. Making sense of health and the environment in early modern England. In *Medicine in Society: Historical Essays*, edited by Andrew Wear. Cambridge: Cambridge University Press.
Wear, Andrew. Fall 2008. Place, health, and disease: The airs, waters, places tradition in early modern England and North America. *Journal for Medieval and Early Modern Studies* 38 (3): 443–466.
Wear, Andrew. 2000. *Knowledge & practice in English Medicine, 1550–1680.* Cambridge: Cambridge University Press.
Withey, Alun. 2015. *Technology, self-fashioning and politeness in eighteenth-century Britain: Refined bodies.* London: Palgrave Macmillan.
Wild, Wayne. 2006. Medicine-by-Post: The changing voice of illness in eighteenth-century British consultation letters and literature. *Clio Medica* 79 (1): 1–286.
Williams, Elizabeth E. 2012. Sciences of appetite in the enlightenment, 1750–1800. *Studies in History and Philosophy of Biological and Biomedical Sciences* 43:392–404.
Williams, Elizabeth. 2020. Eating after the climacteric: Food, gender and ageing in the long eighteenth century. In *Lifestyle and medicine in the enlightenment*, edited by James Kennaway and Rina Knoeff, 67–85. London: Routledge. https://doi.org/10.4324/9780429465642.
Williams, James. September 2008. Sport and the elite in early Modern England. *Sport in History* 28 (3): 389–413.
Wolfe, Charles T. 2021. Vitalism and the metaphysics of life. In *Life and death in early modern philosophy*, edited by Susan James. Oxford: Oxford University Press. https://doi.org/10.1093/oso/9780192843616.003.0016.
Ylivuori, Solle. 2018. *Women and politeness in eighteenth-century England: Bodies, identities and power.* London: Routledge.

Index[1]

[1] Note: Page numbers followed by 'n' refer to notes.

M. Wübbels, *Quantification and Measurement in British Health Advice, c. 1670–1830*, Palgrave Studies in Medieval and Early Modern Medicine,
https://doi.org/10.1007/978-3-032-32760-4

FSC
www.fsc.org
MIX
Papier | Fördert gute Waldnutzung
FSC® C083411

Kolibri 360 GmbH
Ferdinand-Jühlke-Straße 7
99095 Erfurt, Deutschland
produktsicherheit@kolibri360.de